Transition Metals in the Synthesis of Complex Organic Molecules

LOUIS S. HEGEDUS

Transition Metals
in the Synthesis
of Complex
Organic Molecules

Transition Metals in the Synthesis of Complex Organic Molecules

Louis S. Hegedus

Colorado State University

UNIVERSITY SCIENCE BOOKS
Mill Valley, California

University Science Books
20 Edgehill Road
Mill Valley, CA 94941

Fax (415) 383-3167

Library of Congress Catalog Number: 93-085291
ISBN 0-935702-28-8

Printed in the United States of America
10 9 8 7 6 5 4 3 2 1

Contents

Preface

This book is based on short courses presented at pharmaceutical companies, and on a graduate course taught recently at Colorado State University. It is intended for use in a one-semester course to upper-level students familiar with basic organic chemistry, or as a survey course for practicing organic chemists to acquaint them with the many ways transition metal organometallic chemistry might be of use in solving their synthetic problems.

This project was initially intended as the third edition of a graduate text on organotransition metal chemistry, written by James Collman, Jack Norton, Rick Finke, and myself.* It proved impossible, however, to find a willing coauthor for this volume (some people actually learn from experience). Therefore, the chapters dealing with organic synthetic applications (chapters 13–20) were updated through January 1993 and expanded, a brief synopsis of the formalisms and mechanisms was appended, and this volume was prepared as an independent book. Most of the specific examples and applications presented are new, and there is only minimal overlap with the corresponding chapters in the second edition.

Chapters 1 and 2 of this volume present a very brief overview of the formalisms and mechanisms required to understand the processes discussed in the rest of the book. The treatment is minimal and many details and subtleties are not discussed. "Collman, Hegedus, Norton and Finke" remains the source for these details as well as the source for extensive literature references to this material.

Chapters 3–10 deal with the applications of transition metal organometallic chemistry to organic synthesis with specific attention to applications with complex molecules. In addition to "classic" homogeneous hydrogenation, chapter 3 treats the more recent enantioselective reductions

---

*Collman et al., *Principles and Applications of Organotransition Metal Chemistry*, 2nd Edition, University Science Books, 1987, 989 pp.

with BINAP catalysts. Chapter 4, dealing with the chemistry of metal-carbon σ-bonds, represents one of the very active current areas for research, and covers the topics of functionalized organocuprate chemistry, as well as the rapidly growing area of the use of palladium in organic synthesis. Not a lot has happened in the area of carbonylation/decarbonylation chemistry, so chapter 5 represents only a modest update of this area. In contrast, metal carbene chemistry has undergone rapid growth both in the thermal and photochemical reactions of Fischer carbene complexes, and the metal-catalyzed reactions of diazo compounds, all of which receive up to date coverage in chapter 6. Chapter 7 deals with the chemistry of alkene, diene and dienyl complexes, and much of the material in this chapter is new. Both the Pauson-Khand reaction and the Nicholas reaction have proven their worth in the synthesis of complex molecules and chapter 8 presents a summary of these advances, as well as preliminary results on zirconabenzyne complexes. In contrast, cyclotrimerization of alkynes has seen little recent activity, and the material in chapter 8 on this topic has undergone only minor revision. π-Allyl palladium complexes continue to be used in complex syntheses and chapter 9 presents the current state of this area, as well as the chemistry of other π-allyl metal complexes. Finally, chapter 10 deals with π-arene chemistry including the recent advances in the use of these complexes in the control of stereochemistry in reactions of aromatic compounds. "Collman, Hegedus, Norton and Finke" covered the literature through 1985. In contrast, a majority of references in this book are from 1986–1992.

The manuscript was read for clarity and accuracy by Dr. Kathleen Zandi and Professor Ernst Peter Kündig. In addition Professor Kündig provided extensive technical assistance with much of the manuscript, and along with A. Ripa, helped modernize chapter 10. Their many helpful comments greatly improved the manuscript. What errors and inaccuracies remain are solely my responsibility. The manuscript was typed in Microsoft Word 5.0, and the figures and equations executed in ChemDraw 3.0.2 by Michelle Pinkerton.

Louis S. Hegedus
Colorado State University

CHAPTER 1

Formalisms, Electron Counting, Bonding (How Things Work)

1.1 Introduction

Why should an organic chemist even consider using transition metals in complex syntheses? There are many reasons. Virtually every organic functional group will coordinate to some transition metal, and upon coordination, the reactivity of that functional group is often dramatically altered. Electrophilic species can become nucleophilic and vice versa, stable compounds can become reactive and highly reactive compounds can become stabilized. Normal reactivity patterns of functional groups can be inverted, and unconventional (*impossible,* under "normal" conditions) transformations can be achieved with facility. Highly reactive, normally unavailable reaction intermediates can be generated, stabilized, and used as efficient reagents in organic synthesis. Most organometallic reactions are highly specific, able to discriminate between structurally similar sites, thus reducing the need for bothersome "protection-deprotection" sequences that plague conventional organic synthesis. Finally, by careful selection of substrate and metal, multistep cascade sequences can be generated to form several bonds in a single process in which the metal "stitches together" the substrate.

However wonderful this sounds, there are a number of disadvantages as well. The biggest one is that the use of transition metals in organic synthesis requires a new way of thinking, as well as a rudimentary knowledge of how transition metals behave. In contrast to carbon, metals have a number of stable, accessible oxidation states, geometries, and coordination numbers, and their reactivity towards organic substrates is directly related to these features. The structural complexity of the transition metal species involved is, at first, disconcerting. Luckily, a few, easily mastered formalisms provide a logical framework upon which to organize and systematize this large amount of information, and this chapter is designed to provide this mastery.

The high specificity, cited as an advantage above, is also a disadvantage, in that specific reactions are not very general, so that small changes in the substrate can turn an efficient reaction into one that does not proceed at all. However, most transition metal systems carry "spectator" ligands — ligands which are coordinated to the metal but are not directly involved in the reaction — in addition to the organic substrate of interest, and these allow for the fine tuning of reactivity, and an increase in the scope of the reaction.

Finally, organometallic mechanisms are often complex and poorly understood, making both the prediction and the rationalization of the outcome of a reaction difficult. Frequently, a manifold of different reaction pathways are very similar in energy, and seemingly minor changes in the features of a reaction cause it to take an entirely unexpected course. By viewing this as an opportunity to develop new reaction chemistry rather than an obstacle to performing the desired reaction, progress can be made.

1.2 Formalisms

Every reaction presented in this book proceeds in the coordination sphere of a transition metal, and it is the precise electronic nature of the metal that determines the course and the outcome of the reaction. Thus, a very clear view of the nature of the metal is critical to an understanding of its reactivity. The main features of interest are (1) the oxidation state of the metal, (2) the number of d electrons on the metal in the oxidation state under consideration, (3) the coordination number of the metal, and (4) the availability (or lack thereof) of vacant coordination sites on the metal. The simple formalisms presented below permit the easy determination of each of these characteristics. A caveat, however, is in order. These formalisms are just that — formalisms — not reality, not the "truth", and in some cases, not even chemically reasonable. However, by placing the entire organic chemistry of transition metals within a single formalistic framework, an enormous amount of disparate chemistry can be systematized and organized, and a more nearly coherent view of the field is thus available. As long as it remains consistently clear that we are dealing with formalisms, exceptions will not cause problems.

a. Oxidation State

The oxidation state of a metal is defined as the charge left on the metal atom after *all* ligands have been removed *in their normal, closed-shell, configuration* — that is, *with their electron pairs*. The oxidation state is *not* a physical property of the metal, and it cannot be measured. It is a formalism which helps us in counting electrons, but no more. Typical examples are shown in Figure 1.1. (An alternative formalism, common in the older literature and still prevalent among Europeans, removes each covalent, anionic ligand as a *neutral* species with a single electron (homolytically), leaving the other electron on the metal. Although this results in a different *formal* oxidation state for the metal, it leads to the same conclusions as to the total number of electrons in the bonding shell, and the degree of coordination saturation.)

It must be emphasized that the *chemical* properties of the ligands are not always consonant with the oxidation state formalism. In metal hydrides, the hydride ligand is *always* formally considered to be H^-, even though some transition metal "hydrides" are strong acids! Despite this, the formalism is still useful.

Figure 1.1   Examples of Oxidation State Determination

b. *d* Electron Configuration, Coordination Saturation and the 18 Electron Rule

Having assigned the oxidation state of the metal in a complex, the number of *d* electrons on the metal can easily be assessed by referring to the periodic table. Figure 1.2 presents the transition elements along with their *d*-electron count. The transition series is formed by the systematic filling of the *d* orbitals. Note that these electron configurations differ from those presented in most elementary texts in which the 4*s* level is presumed to be lower in energy than the 3*d*, and is filled first. Although this is the case for the *free atom* in the elemental state, these two levels are quite close in energy, and, for the *complexes* — which are not free metal atoms, but rather metals surrounded by ligands — discussed in this text, the assumption that the outer electrons are *d* electrons is a good approximation. By referring to the periodic table (or preferably, by re-membering the positions of the transition metals) the *d* electron count for any transition metal in any oxidation state is easily found.

The *d* electron count is critical to an understanding of transition metal organometallic chemistry because of the **18 electron rule**, which states "in mononuclear, diamagnetic complexes, the total number of elec-

Group number (new)		4	5	6	7	8	9	10	11
First row	3d	Ti	V	Cr	Mn	Fe	Co	Ni	Cu
Second row	4d	Zr	Nb	Mo	Tc	Ru	Rh	Pd	Ag
Third row	5d	Hf	Ta	W	Re	Os	Ir	Pt	Au
	0	4	5	6	7	8	9	10	---
	I	3	4	5	6	7	8	9	10
Oxidation	II	2	3	4	5	6	7	8	9
state	III	1	2	3	4	5	6	7	8
	IV	0	1	2	3	4	5	6	7

$\left. \right\} d^n$

Figure 1.2 d-Electron Configuration for the Transition Metals as a Function of Formal Oxidation State

trons in the bonding shell (the sum of the metal d electrons plus those contributed by the ligands) never exceeds 18" (at least not for very long — see below). This 18 electron rule determines the *maximum* allowable number of ligands for any transition metal in any oxidation state. Compounds having the maximum allowable number of ligands — having 18 electrons in the bonding shell — are said to be coordinatively saturated — that is, there are *no* remaining coordination sites on the metal. Complexes *not* having the maximum number of ligands allowed by the 18 electron rule are said to be coordinatively unsaturated — that is, they have vacant coordination sites. Since vacant sites are usually required for catalytic processes (the substrate must coordinate before it can react), the degree of coordination is central to many of the reactions presented below.

c. Classes of Ligands

The very large number of ligands that are involved in organotransition metal chemistry can be classified into three families: (1) *formal* anions, (2) *formal* neutrals, (3) *formal* cations. These families result from the oxidation state formalism requiring the removal of ligands in their closed-shell (with their pair(s) of electrons) state. Depending on the ligand, it can be either electron donating or electron withdrawing, and its specific nature has a profound effect on the reactivity of the metal center. Ligands with additional unsaturation may coordinate to more than one site, and thus contribute more than two electrons to the total electron count. Examples of each of these will be presented below, listed in approximate order of decreasing donor ability.

Formal anionic ligands which act as two electron donors are:

$$R^- > Ar^- > H^- > R\overset{\displaystyle O}{\overset{\|}{C}}(\text{-}) > X^- \text{ where } X^- = \text{halogen, CN}^-$$

These ligands fill one coordination site through one point of attachment and are called "monohapto" ligands, designated as η^1. The allyl group, $C_3H_5^-$, can act as a monohapto (η^1) two electron donor, or a trihapto (η^3) (π-allyl) four electron donor (Figure 1.3). In the latter case, *two* coordination sites are filled, and all three carbons are bonded to the metal, but the ligand as a whole is still a formal mono anion. The cyclopentadienyl ligand, $C_5H_5^-$, most commonly bonds in an η^5 fashion, filling three coordination sites and acting as a six electron donor, although η^3 (four electrons, two site, equivalent to η^3-allyl) and η^1 coordinations are known. The cyclohexadienyl ligand, produced from nucleophilic attack on η^6-arene metal complexes (Chapter 10), is almost invariably a six electron, mononegative, η^5-ligand which fills three coordination sites. These ligands are illustrated in Figure 1.3.

Formal neutral ligands abound, and they encompass not only important classes of "spectator" ligands — ligands such as phosphines and amines introduced to moderate the reactivity of the metal, but which are not *directly* involved in the reaction under consideration — but also ligands such as carbon monoxide, alkenes, alkynes, and arenes, which are often *substrates* (Figure 1.4). Ligands such as phosphines and amines are good σ donors in organometallic reactions, and increase the electron density at the metal, while ligands such as carbon monoxide, isonitriles, and olefins are π acceptors, and decrease the electron density at the metal. The reason for this is presented in the next section.

Figure 1.3   Bonding Modes for Mononegative Anionic Ligands

Figure 1.4   Bonding Modes for Neutral Ligands

Formally cationic ligands are much less common, since species that bear a full formal positive charge *and* a lone pair of electrons are rare. The nitrosyl group is one of these, being a cationic two electron donor. It is often used as a spectator ligand, or to replace a carbon monoxide, to convert a neutral carbonyl complex to a cationic nitrosyl complex.

With all of the above information in hand, it is now possible to consider virtually any transition metal complex, assign the oxidation state of the metal, assess the total number of electrons in the bonding shell, and decide if that complex is coordinatively saturated or unsaturated. For example, the complex $CpFe(CO)_2(C_3H_7)$ is a stable, neutral species, containing two formally mononegative ligands, the n-propyl (C_3H_7) group, an η^1, $2e^-$ donor ligand, and the Cp ligand, an η^3 $6e^-$ donor ligand, and two neutral carbon monoxide ligands which contribute two electrons each. Since the overall complex is neutral, and has two mononegative ligands, the iron must have a formal 2+ charge, making it Fe^{II}, d^6. For an overall electron count, there are six electrons from the metal, a total of four from the two CO's, two electrons from the propyl group and six electrons from the Cp group, for a grand total of $18e^-$. Thus, this complex is an Fe^{II}, d^6, coordinatively saturated complex. Other examples, with explanatory comments, are given in Figure 1.5.

The last example in Figure 1.5 illustrates one of the difficulties in the strict application of formalisms. Strictly speaking, this treatment of the cyclohexadienyliron tricarbonyl cation is the formally correct one, since the rules state that each ligand shall be removed *in its closed shell form*, making the cyclohexadienyl ligand a six electron donor *anion*. However, this particular complex is prepared by a *hydride* abstraction from the neutral cyclohexadieneiron tricarbonyl, making the assumption that the cyclohexadienyl ligand is four electron donor cation reasonable. Further, the cyclohexadienyl group in the complex is quite reactive towards nucleophilic attack (Chapter 7). The true situation is impossible to establish. In reality, the plus charge resides neither exclusively on the metal, as the first, formally correct, treatment would assume, nor completely on the cyclohexadienyl ligand from which the hydride was abstracted, but

Figure 1.5   Electron Counting, Oxidation State

5 x CO, 2 e⁻ each, no charge | Complex is neutral, no charged ligands

, 2 e⁻, no charge

$\therefore$ Cr⁰, d^6

electron count $d^6 = 6e^-$ 10e⁻ from

5 CO's; 2e⁻ from :C⟨

_____

18e⁻ saturated

6e⁻, no charge | Complex is neutral, no charged ligands

3 x CO 2e⁻ each, no charge

$\therefore$ Cr⁰, d^6

electron count $d^6 = 6e^-$ ⬡ = 6e⁻

3 x CO = 6e⁻

_____

18e⁻ saturated

For each Pd

allyl 4e⁻, −1 charge | Complex is neutral, two −1 ligands

Cl⁻ 2e⁻, −1 charge

$\therefore$ Pd^II, d^8

Cl → 2e⁻, no charge

electron count $d^8 = 8e^-$;

allyl = 4e⁻, 2 Cl's = 4e⁻

_____

16e⁻ unsaturated

bridging by lone pair on Cl
each Cl acts as a 2e⁻, mono negative
ligand to one of the Pd's, and a 2e⁻
neutral donor ligand (like phosphine) to
the other

6e⁻, −1 charge | Complex is +1 charged, one −1 ligand

3 x CO 2e⁻ each, no charge

$\therefore$ Fe^II, d^6

electron count $d^6 = 6e^-$

3 x CO = 6e⁻

cyclohexadienyl = 6e⁻

_____

18e⁻ saturated

Figure 1.5 Continued

rather is distributed throughout the entire metal-ligand array. Thus *neither* treatment represents the true situation, while *both* treatments come to the conclusion that the complex is an 18e$^-$ saturated complex. The power of the formalistic treatment is that it is consistent, and does *not* require a knowledge of how the complex is synthesized or how it reacts. As long as formalisms are not taken as a serious representation of reality, but rather treated as a convenient way of organizing and unifying a vast amount of information, they will be useful. To quote Roald Hoffmann "Formalisms are convenient fictions which contain a piece of the truth . . . and it is so sad that people spend a lot of time arguing about the deductions they draw . . . from formalisms without worrying about their underlying assumptions."[1]

1.3   Bonding Considerations[2]

The transition series is formed by the systematic filling of the *d* shell which, in complexes, is of lower energy than the next *s, p* level. Thus, transition metals have partially filled *d* orbitals, and vacant *s* and *p* orbitals. In contrast, most of the ligands have filled "*sp^n*" hybrid orbitals, and, for unsaturated organic ligands, vacant anti-bonding π^* orbitals. The *d* orbitals on the metal have the same symmetry *and* similar energy as the antibonding π^* orbitals of the unsaturated ligands (Figure 1.6). Transition metals participate in two types of bonding, often simultaneously. σ-*Donor* bonds are formed by the overlap of filled *sp^n* hybrid orbitals on the ligand (including π-bonding orbitals of unsaturated hydrocarbons) with vacant "*dsp*" hybrid orbitals on the metal. Ligands which are primarily σ-donors, such as R_3P, R_3N, H^-, and R^-, *increase* the electron density on the metal. Unsaturated organic ligands, such as alkenes, alkynes, arenes, carbon monoxide and isonitriles, which have π^* antibonding orbitals, bond somewhat differently. They, too, form σ-donor bonds by overlap of their filled π-bonding orbitals with the vacant "*dsp*" hybrid orbitals of the metal. In addition, *filled* d *orbitals* of the metal can overlap with the *vacant* π^*-antibonding orbital of the ligand, back donating

d-orbitals   π* orbital

Figure 1.6   Symmetry of _d_ Orbitals and π* Orbitals

Organic π-bond as a σ-donor   Organic π*-orbital as π-acceptor

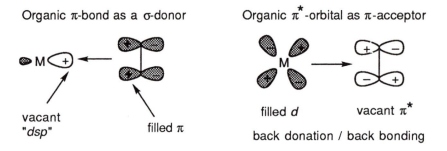

vacant   filled π   filled _d_   vacant π*
"_dsp_"

back donation / back bonding

Figure 1.7   Details of π-Bonding

electron density from the metal to the ligand (Figure 1.7). Thus, unsaturated organic ligands can act as π-acceptors or π-acids, _decreasing_ electron density on the metal. Because of these two modes of bonding — σ-donation and π-accepting "back bonding" — metals can act as an electron sink for ligands, either supplying or accepting electron density. As a consequence, the electron density on the metal, and hence its reactivity, can be modulated by varying the ligands around the metal. This offers a major method for fine-tuning the reactivity of organometallic reagents.

There are two modes of π-back bonding and two types of π-acceptor ligands: (1) longitudinal acceptors, such as carbon monoxide, isonitriles and linear nitrosyls, and (2) perpendicular acceptors, such as alkenes and alkynes (Figure 1.8).

There is ample physical evidence for π-back bonding with good π-acceptor ligands. Carbon monoxide, which always acts as an acceptor, experiences a lengthening of the CO bond and a decrease in CO stretching frequency in the infrared spectrum upon complexation, both indications of the population of a π* orbital on CO, decreasing the CO bond order. The situation is a little more complex with olefins. With electrophilic metals, such a Pd^{2+} or Pt^{2+}, olefins are primarily σ-donor ligands, and the C=C bond length in such complexes is virtually the same as for the free olefin.

Longitudinal   Perpendicular

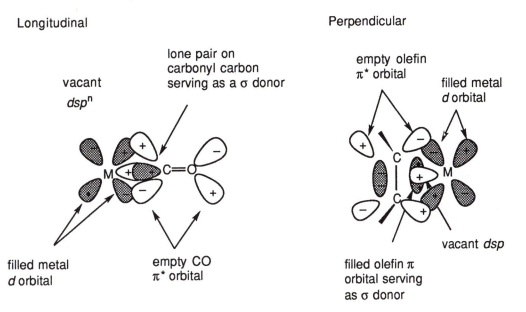

Figure 1.8   Types of π-acceptors

With electron-rich metals, such as Pd(0), substantial back bonding results, the olefin C–C bond is lengthened, and the hybridization at the olefinic carbons changes towards sp^3. The degree of π-back bonding has a serious effect on the reactivity of π-acceptor ligands. Organometallic processes involving reactions of π-acceptor ligands are the next topic for consideration.

1.4   Structural Considerations

Although the focus of this book is the *reactivity* of metal-bound ligands, it is important to appreciate the geometries adopted by organometallic complexes. For 18 electron (saturated) complexes these geometries are determined by steric factors, such that four coordinate complexes such as $Ni(CO)_4$ and $Pd(PPh_3)_4$, are tetrahedral, five coordinate complexes such as $Fe(CO)_5$ and $(diene)Fe(CO)_3$, are trigonal bipyramidal, and six coordinate complexes such as $Cr(CO)_6$, Ir^{III} complexes, and Rh^{III} complexes, are octahedral.

In contrast, the d^8 complexes of Pd^{II}, Pt^{II}, Ir^{I}, and Rh^{I} find it energetically favorable to form square planar, four coordinate unsaturated $16e^-$ complexes. The planar geometry makes the $d_{x^2-y^2}$ orbital so high in energy that it remains unoccupied in stable complexes, but enables such complexes to undergo facile ligand substitution by an associative mechanism. This, and other mechanistic considerations are the topic of the next chapter.

References

1. Sailland, J-V.; Hoffmann, R. *J. Am. Chem. Soc.* **1984**, *106*, 2006.
2. For a more rigorous treatment of bonding see: Collman, J.P.; Hegedus, L.S.; Finke, R.O.; Norton, J.R. *Principles and Applications of Organotransition Metal Chemistry*. University Science Books, Mill Valley, CA, 1987, pp. 21-56.

CHAPTER 2

Organometallic Reaction Mechanisms

2.1   Introduction

The focus of this entire book is the *organic* chemistry of organometallic complexes, and the effect coordination to a transition metal has on the reactivity of organic substrates. In transition metal catalyzed reactions of organic substrates, it is the chemistry of the *metal* that determines the course of the reaction, and a rudimentary understanding of the mechanisms by which transition metal organometallic complexes react is critical to the rational use of transition metals in organic synthesis.

Mechanistic organometallic chemistry is both more complex and less developed than mechanistic organic chemistry, although many parallels exist. The organic chemist's reliance on "arrow pushing" to rationalize the course of a reaction is equally valid with organometallic processes, provided the appropriate rules are followed. Some organometallic mechanisms are understood in exquisite detail, while others are not understood at all. The field is very active, and new insights abound. However, this book will treat the topic only at a level required to permit the planning and execution of organometallic processes, and to facilitate the understanding of the literature in the field. More specialized treatises should be consulted for a more thorough treatment.[1]

2.2 Ligand Substitution Processes

Ligand substitution (exchange) processes (Eq. 2.1) are central to virtually all organometallic reactions of significance for organic synthesis. For catalytic processes, it is common that a stable catalyst precursor must lose a ligand, coordinate the substrate, promote whatever reaction is being catalyzed and then release the substrate. Three of the four steps are ligand exchange processes. In other instances, it may be necessary to replace an innocent (spectator) ligand in an organometallic complex to adjust the reactivity of a coordinated substrate (Eq. 2.2) or to stabilize an unstable intermediate for isolation or mechanistic studies (Eq. 2.3). Thus, a fundamental understanding of ligand exchange processes is central to the utilization of organometallic complexes in synthesis.

Eq. 2.1

$$M-L \ + \ L' \ \rightleftharpoons \ M-L' \ + \ L$$

Eq. 2.2

unreactive toward
nucleophiles

reactive toward
nucleophiles

Eq. 2.3

requires HCl/HNO₃
for oxidation

unstable
to air

requires HNO₃
for oxidation

Figure 2.1   Classification of Ligand Substitution Processes

Ligand exchange processes at metal centers share many common characteristics with organic nucleophilic displacement reactions at carbon centers, in that the outgoing ligand departs with its pair of electrons, while the incoming ligand attacks the metal with *its* pair of electrons. An obvious difference is that metals can have coordination numbers larger than four, and stable, coordinatively unsaturated metal complexes are common. Thus, the *details* of ligand exchange are somewhat different, and more varied, than typical organic nucleophilic displacement reactions. However ligand substitution processes can be classified in much the same way as organic nucleophilic displacement reactions, and can involve either two electron or one electron processes, in associative (S_N2-like) or dissociative (S_N1-like) reactions (Figure 2.1).

By far the most extensively studied ligand exchange reactions are those of coordinatively unsaturated, 16 electron, d^8, square planar complexes of Ni(II), Pd(II), Pt(II), Rh(I) and Ir(I). These typically involve two electron, associative processes, which resemble S_N2 reactions, but with some differences. The platinum system shown in Eq. 2.4 is typical.[2]

Eq. 2.4

SₙN2 - apical attack   sq. py.   trig. bip.

sq. py.

Because the metal is coordinatively unsaturated, apical attack at the vacant site occurs, producing a square pyramidal, saturated intermediate. This rearranges via a trigonal bipyramidal intermediate (analogous to the transition state in organic S_N2 reactions) to the square pyramidal intermediate with the leaving group axial; the leaving group then leaves. The rate is second order, and depends upon the metal, in the order Ni > Pd >> Pt, with a range of ~10^6. This is thought to reflect the relative abilities of the metals to form five-coordinate, 18 electron intermediates. As a consequence of this rate difference, if a reaction depends on ligand exchange, platinum-catalyzed processes may be too slow to be synthetically useful. The rate also depends on the incoming ligand, Y, in the order R_3P > Py > NH_3, Cl^- > H_2O > OH^-, spanning a range of ~10^5. Further the rate depends on the leaving group X, in the order NO_3^- > H_2O > Cl^- > Br^- > I^- > N_3^- > SCN^- > NO_2^- > CN^-. The rate depends on the ligand *trans* to the ligand being displaced, in the order R_3Si^- > H^- ≈ CH_3^- ≈ CN^- ≈ olefins ≈ CO > PR_3 ≈ NO_2^- ≈ I^- ≈ SCN^- > Br^- > Cl^- > RNH_2 ≈ NH_3 > OH^- > NO_3^- ≈ H_2O. A practical consequence of this "*trans* effect" is the ability to activate ligand exchange by replacing the ligand *trans* to the one to be displaced.

In contrast, 18 electron *saturated* systems undergo ligand exchange reactions much more slowly, and then by a dissociative, S_N1-like process. That is, a coordination site must be vacated before ligand exchange can occur. Typical complexes which undergo ligand exchange by this process are the saturated metal carbonyl complexes, such as $Ni(CO)_4$ ($M°$, d^{10}, saturated), $Fe(CO)_5$ ($M°$, d^8, saturated) and $Cr(CO)_6$ ($M°$, d^6, saturated). Nickel tetracarbonyl is the most labile and most reactive toward ligand exchange (Eq. 2.5). The rate of exchange is first order, and proportional to the concentration of nickel carbonyl. With ligands other than CO, the rate of ligand exchange can be accelerated by bulky ligands, since loss of one ligand leads to release of steric strain. This is best illustrated by the nickel(0) (tetrakis) phosphine complexes seen in Table 2.1. A measure of the bulk of a phosphine is its "cone angle" (Figure 2.2).[3] As the size of the phosphine increases, the equilibrium constant for loss of a ligand changes by more than 10^{10}. A practical consequence is that, in catalytic processes involving loss of a phosphine ligand, reactivity can be "fine tuned" by alteration of the phosphine ligand.

Eq. 2.5

$$Ni(CO)_4 \underset{}{\overset{slow}{\rightleftharpoons}} Ni(CO)_3 + CO \xrightarrow[\text{fast}]{L} LNi(CO)_3$$

tetrahedral
labile

Rate α [$Ni(CO)_4$] - 1st order

Table 2.1   Ligand Dissociation as a Function of Phosphine Cone Angles

L =	P(OEt)$_3$	P(O-*p*-tolyl)$_3$	P(O-*i*Pr)$_3$	P(O-*o*-tolyl)$_3$	PPh$_3$
Cone angle	109°	128°	130°	141°	145°
K_D	<10^{-10}	6 x 10^{-10}	2.7 x 10^{-5}	4 x 10^{-2}	No NiL$_4$

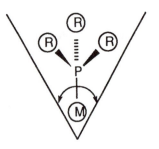

Figure 2.2   Cone Angle

In contrast to nickel tetracarbonyl, iron pentacarbonyl (trigonal bipyramidal) and chromium hexacarbonyl (octahedral) are *not* labile, and are quite inert to ligand substitution, and either heat, sonication, or irradiation (light) is required to affect ligand substitution in these systems. Carbon monoxide displacement in these substitutionally inert complexes can also be chemically promoted. Tertiary amine oxides irreversibly oxidize metal-bound carbon monoxide to carbon dioxide, which readily dissociates. Since amines are not strong ligands for metals in low oxidation states, a coordination site is opened (Eq. 2.6).[4] Amine oxides are not particularly powerful oxidants for organic compounds, and have been used to remove metal carbonyl fragments — by "chewing off" the CO's— from organic ligands reluctant to leave on their own. This has been used extensively in organic synthetic applications of cationic dienyliron carbonyl complexes (Chapter 7). Tributylphosphine oxide can *catalyze* CO exchange, presumably by a similar mechanism (Eq. 2.7),[5] although in this case the strength of the P–O bond prevents fragmentation and oxidation of CO.

Eq. 2.6

Eq. 2.7

Ligand exchanges can also be promoted by one-electron oxidation or reduction of 18 e⁻ systems, transiently generating the much more labile 17 e⁻ or 19 e⁻ systems. However, to date these processes have found little application in the use of transition metals in organic synthesis.

Not all 18e⁻ complexes undergo ligand substitution by a dissociative mechanism. In these cases, an exchange phenomenon called "slippage" is important. Ligands such as η^3-allyl, η^5-cyclopentadienyl, and η^6 arene normally fill multiple coordination sites. However, even in saturated compounds, they can open coordination sites on the metal by "slipping" to a lower hapticity (coordination number). This is illustrated in Figure 2.3. This "slippage" can account for a number of otherwise difficult to explain observations, including apparent associative S_N2-like ligand exchange with coordinatively saturated complexes (for example, Eq. 2.8). With arene complexes in particular, the first step ($\eta^6 \rightarrow \eta^4$) requires the most energy, since aromaticity is disrupted. With fused arene ligands such as naphthalene or indene, this step is easier, and rate enhancements of 10^3-10^8 have been observed.

Eq. 2.8

η³-allyl   η¹-allyl   η⁵ Cp   η³ Cp   η¹ Cp

(2 sites)   (1 site)   (3 sites)   (2 sites)   (1 site)

η⁶-arene   η⁴-arene   η²-arene

(3 sites)   (2 sites)   (1 site)

Figure 2.3   Slippage in Polyhapto Complexes

2.3   Oxidative Addition/Reductive Elimination[6]

Oxidative addition/reductive elimination processes are central to a vast array of synthetically useful organometallic reactions (e.g., the "Heck" reaction, Chapter 4), and occur because of the ability of transition metals to exist in several different oxidation states, in contrast to nontransition metal compounds, which usually have closed shell configurations. In fact, it is this facile shuttling between oxidation states that makes transition metals so useful in organic synthesis. The terms oxidative addition and reductive elimination are generic, describing an overall transformation but not the specific mechanism by which that transformation occurs. There are many mechanisms for oxidative addition, some of which are presented below. The general transformation is described in Eq. 2.9.

A coordinatively unsaturated transition metal, usually in a low (0, +1) oxidation state and hence relatively electron rich and nucleophilic, undergoes reaction with some substrate A–B, to form a new complex A–M–B in which the metal has formally inserted into the A–B bond. Because of the oxidation state formalism, which demands each ligand be removed with its pair of electrons (although A and B really contribute only one electron each to A–M–B), the metal is formally oxidized in the A–M–B adduct. Since the "oxidant" A–B has added to the metal, the process is termed "oxidative addition." In general, ligands such as R_3P, R^- and H^-, which are good σ-donors and increase the electron density at the metal, facilitate oxidative addition, whereas ligands such as CO, CN^- and olefins, which are good π-acceptors and decrease electron density at the metal, suppress oxidative addition. The reverse of oxidative addition is reductive elimination, and both the transformation and terminology are obvious. Oxidative addition-reductive elimination processes are central to the use of transition metals in organic synthesis because one *need not reductively eliminate the same two groups that were oxidatively added to the metal*. This is the basis of a very large number of cross-coupling reactions (Chapter 4).

Eq. 2.9

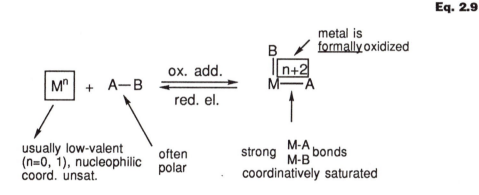

The most commonly encountered systems are Ni(0), Pd(0) → Ni(II), Pd(II) ($d^{10} \rightarrow d^8$) and Rh(I), Ir(I) → Rh(III), Ir(III) ($d^8 \rightarrow d^6$). The most extensively studied system and the one that contributed heavily to the development of many of the formalisms presented above is Vaska's compound, the lemon yellow, Ir(I), d^8, 16 electron, coordinatively unsaturated complex which undergoes oxidative addition with a wide array of substrates to form stable Ir(III), d^6, 18 electron saturated complexes (Eq. 2.10).

Eq. 2.10

The range of addenda that participate in oxidative addition reactions to transition metal complexes is large, and contains many members useful in organic synthesis. There are three general classes of addend: polar electrophiles, nonpolar substrates and multiple bonds (Figure 2.4). Note that many of these result in the formation of metal-hydrogen and metal-carbon bonds, and in the "activation" of a range of organic substrates.

There are several documented mechanisms for oxidative addition, and the one followed depends heavily on the nature of the reacting partners. They include (1) concerted, associative, one step "insertions" into A–B by M, (2) ionic, associative, two step S_N2 reactions, and (3) electron-transfer - radical chain mechanisms.

Concerted oxidative additions are best known for nonpolar substrates, particularly the oxidative addition of H_2 (central to catalytic hydrogenation), and oxidative addition into C–H bonds (hydrocarbon activation). The process is thought to involve prior coordination of the H–H or C–H bond to the metal in an "agostic" (formally a two electron, three center bond) fashion, followed by *cis* insertion (Eq. 2.11). This idea is supported by the recent isolation of molecular hydrogen complexes of tungsten (Eq. 2.12)[7] in which just this sort of bonding is observed. This

a. Polar electrophiles: X_2, HX, R—X, RC(=O)—X, RSO_2–X

b. Nonpolar: H_2, R_3Si–H, RC(=O)—H, R—H, Ar—H

c. Multiple bonds (A-B stays connected): O_2, S_2

Figure 2.4   Classes of Substrates for Oxidative Addition Reactions

same process is likely but not yet demonstrated for oxidative addition of M(0) d^{10} complexes with sp^2 halides, a process that goes with retention of stereochemistry at the olefin (Eq. 2.13).

Eq. 2.11

"agostic"

Eq. 2.12

Eq. 2.13

Oxidative addition via S_N2-type processes is most often observed with strongly nucleophilic, low valent metals and classic S_N2 substrates such as primary and secondary organic halides and tosylates. An early example is shown in Eq. 2.14,[8] and its features are typical of metals which react by S_N2-type processes. The rate law is second order, first order metal and first order in substrate, and polar solvents accelerate the reaction, as expected for polar intermediates. The order of substrate reactivity is I > Br > Cl as expected for S_N2-type reactions. The rate is also dependent on the nature of the spectator phosphine ligands, and, in general, increases with increasing basicity (ability to donate electrons to the metal) of the phosphine. This is an important observation, since it confirms the claim that reactivity of transition metal complexes can be "fine tuned" by changing the nature of the spectator ligands. This feature is exceptionally valuable when using transition metals in complex total synthesis, since the inherent specificity of organometallic reactions can be applied over a wide range of substrates.

Eq. 2.14

($Ir^{(I)}$, d^8, 16 e$^-$)   ($Ir^{(III)}$, d^6, 18e$^-$)

X = I > Br > Cl

L = Et_3P > Et_2PPh > $EtPPh_2$ > PPh_3 (30 fold rate difference)

A synthetically more useful complex that reacts by an S_N2-type mechanism is Collman's reagent, $Na_2Fe(CO)_4$ (Eq. 2.15).[9] This very potent nucleophile is prepared by reducing the iron(0) complex $Fe(CO)_5$ with sodium-benzophenone ketyl to produce the iron(–2) complex. This is an un-usually low oxidation state for iron, but, because carbon monoxide ligands are powerful π-acceptors and strongly electron withdrawing, they stabi-lize negative charge, and highly reduced metal carbonyl complexes are common. The reaction of this complex with halides has all the features common to S_N2 chemistry (Eq. 2.15), and the S_N2 mechanism is the most likely pathway for the reaction of d^8–d^{10} transition metal complexes with 1°, 2°, allylic and benzylic halides.

Eq. 2.15

$$Na_2Fe(CO)_4 \ + \ RX \longrightarrow [RFe(CO)_4]Na \ + \ NaX$$

Rate $\propto$ [RX][Fe(CO)$_4^{2-}$]   for RX 1° > 2°   X = I > Br > OTs > Cl

Stereochemistry at carbon - clean inversion

However, in many cases there is a competing radical chain process which can affect the same overall transformation, but by a completely different mechanism. For example, Vaska's complex (Eq. 2.14) is unreac-tive towards organic bromides and even 2° iodides when the reaction is carried out with the strict exclusion of air. However, traces of air or addi-tion of radical initiators promotes a smooth reaction with these sub-strates, via a radical chain mechanism (Eq. 2.16).[10] In this case, racemization is observed, as expected. The existence of an accessible radical chain mechanism for oxidative addition explains the empirical observation that traces of air sometimes promote the reaction of complexes which normally require strict exclusion of air, and serves as a warning that the mechanism by which a reaction proceeds may be a function of the care with which it is carried out.

Eq. 2.16

A more synthetically useful example of a radical chain oxidative addition is the nickel catalyzed coupling of aryl halides to biaryls, which was originally thought to proceed by two sequential two-electron oxidative additions. However, careful studies showed this was not the case. Rather, a complicated radical chain process involving Ni(I) → Ni(III) oxidative additions is the true course of the reaction (Eq. 2.17).[11]

Reductive elimination is the reverse of oxidative addition and is usually the last step of many catalytic processes. It is exceptionally important for organic synthetic applications because *it is the major way in which transition metals are used to make carbon-carbon and carbon-hydrogen bonds!* In spite of its importance, reductive elimination has not been extensively studied. It is known that the groups to be eliminated must occupy *cis* positions on the metal, or must rearrange to *cis* if they are *trans* (Eq. 2.18).[12] Reductive elimination can be promoted. Anything which reduces the electron density at the metal facilitates reductive elimination. This can be as simple as the dissociation of a ligand, either spontaneously or by the application of heat or light. Both chemical and electrochemical oxidation promotes reductive elimination (although oxidatively driven reductive elimination sounds like an oxymoron). Of practical use is the observation that the addition of strong π-acceptor ligands, such as carbon monoxide, maleic anhydride, quinones or tetracyanoethylene, promotes reductive elimination.

Eq. 2.17

$$L_nNi + ArX \longrightarrow ArAr + NiX_2$$

$$Ni^0 + ArX \longrightarrow Ar{-}NiX^{II} \xrightarrow{\text{ArX}} \underset{\underset{X}{|}}{\overset{\overset{Ar}{|}}{Ar{-}Ni^{IV}{-}X}} \longrightarrow Ar{-}Ar$$

original (incorrect) mechanism

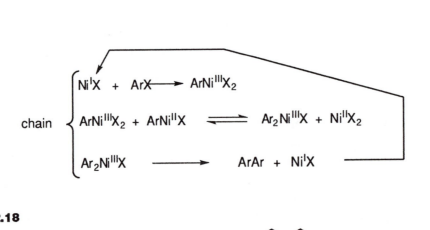

actually a <u>radical chain</u>

initiation

$$ArX + L_3Ni^0 \xrightarrow[\text{ox. add.}]{2e^-} \underset{\underset{L}{|}}{\overset{\overset{L}{|}}{Ar{-}Ni^{II}{-}X}}$$

$$\overset{X}{\underset{}{L_2Ni^{II}Ar}} + ArX \xrightarrow{e^- \text{ transfer}} [ArNi^{III}XL_2]^{\overset{\bullet}{+}}\,[ArX]^{\overset{\bullet}{-}}$$

$$[ArNi^{III}XL_2]^{\overset{\bullet}{+}}\,[ArX]^{\overset{\bullet}{-}} \longrightarrow \boxed{Ni^IX} + \;.\;.\;.\;.\;?$$

$$\underline{or}\; L_nNi^0 + ArX \xrightarrow{e^- \text{ transfer}} [L_nNi\,\overset{\bullet}{+}]\,[ArX]^{\overset{\bullet}{-}} \longrightarrow \boxed{Ni^IX} + Ar^{\bullet} ...$$

chain

$$Ni^IX + ArX \longrightarrow ArNi^{III}X_2$$

$$ArNi^{III}X_2 + ArNi^{II}X \rightleftharpoons Ar_2Ni^{III}X + Ni^{II}X_2$$

$$Ar_2Ni^{III}X \longrightarrow ArAr + Ni^IX$$

Eq. 2.18

2.4 Migratory Insertion / β-Hydride Elimination

The third organometallic mechanism of interest for organic synthesis is the process of migratory insertion (Eq. 2.19), in which an unsaturated ligand — usually CO, RNC, olefins, or alkynes — *formally* inserts into an adjacent (*cis*) metal-ligand bond. When this adjacent ligand is hydride, or σ-alkyl, the process forms a new carbon-hydrogen or carbon-carbon bond, and in the case of alkenes, a new metal-carbon bond. The term "insertion" is somewhat misleading, in that the process actually proceeds by *migration* of the adjacent ligand to the metal bound unsaturated species, generating a vacant site. Insertions are usually reversible. If the migrating group has stereochemistry, it is usually retained in both the insertion step and the reverse.

Eq. 2.19

$$\underset{M-Y}{\overset{X}{\shortmid}} \; \rightleftharpoons \; M-Y-X \; \rightleftharpoons \; \underset{M-Y-X}{\overset{L}{\shortmid}}$$

Y = CO, RNC, C=C, C≡C   X = H, R, etc.

The insertion of carbon monoxide into a metal-carbon σ-bond is one of the most important processes for organic synthesis because it permits the direct introduction of molecular carbon monoxide into organic substrates, to produce aldehydes, ketones, and carboxylic acid derivatives. The reverse process is also important, as it allows for the decarbonylation of aldehydes to alkanes and acid halides to halides.

The general process of CO insertion, shown in Eq. 2.20, involves reversible migration of an R group to a *cis* bound CO, as the rate determining step. The vacant site generated by this migration is then filled by some external ligand present. Similarly, the reverse process requires a vacant coordination site *cis* to the acyl ligand. Both the migratory insertion and the reverse deinsertion occur with retention of configuration of the migrating group and of the metal, if the metal center is chiral. The migratory aptitude is: η^1 allyl ≥ Et > Me > PhCH$_2$ > vinyl ≥ aryl, ROCH$_2$> HOCH$_2$. Hydride (H), CH$_3$CO, and CF$_3$ usually do not migrate, and heteroatoms such as RO$^-$ and R$_2$N$^-$ rarely migrate. Lewis acids often accelerate CO insertion reactions, as does oxidation, although in this case the metal acyl species often continues to react, and is oxidatively cleaved from the metal.

Eq. 2.20

Alkenes also readily insert into metal-hydrogen bonds (a key step in catalytic hydrogenation of olefins) and metal-carbon σ-bonds resulting in alkylation of the olefin (Eq. 2.21). The reverse of hydride insertion is β-hydrogen elimination, a very common pathway for decomposition of s-alkylmetal complexes. Olefin insertions share many features. The olefin *must* be coordinated to the metal prior to insertion, and must be *cis* to the migrating hydride or alkyl group. The migration generates a vacant site on the metal, and for the reverse process, β-hydride elimination (β-alkyl elimination does not occur) a vacant *cis* site on the metal is required. Both the metal and the migrating group add to the same face of the olefin. When an alkyl group migrates, its stereochemistry is maintained. A rough order of migrating aptitude is: H >> R, vinyl, aryl > RCO >> RO, R_2N. Again, heteroatom groups migrate only with difficulty, since the extra lone pair(s) on these ligands can form multiple bonds to the metal (Eq. 2.22).

Eq. 2.21

Eq. 2.22

Alkynes also insert into metal-hydrogen and metal-carbon bonds, again with *cis* stereochemistry and with retention at the migrating center (Eq. 2.23). This has been less studied, and is complicated by the fact that alkynes often insert into the product s-vinyl complex, resulting in oligomerization.

Eq. 2.23

2.5 Nucleophilic Attack on Ligands Coordinated to Transition Metals

The reaction of nucleophiles with coordinated, unsaturated organic ligands is one of the most useful processes for organic synthesis.[13] Unsaturated organic compounds such as carbon monoxide, alkenes, alkynes and arenes are electron rich, and are usually quite unreactive towards nucleophiles. However, complexation to electron deficient metals inverts their normal reactivity, making them generally subject to nucleophilic attack, and opening up entirely new synthetic transformations. The more electron deficient a metal center is, the more reactive towards nucleophiles its associated ligands become. Thus, cationic complexes and complexes having strong π-accepting spectator ligands such as carbon monoxide are especially reactive in these processes.

Metal-bound carbon monoxide is generally reactive towards nucleophiles, providing a major route for the incorporation of carbonyl groups into organic substrates. Neutral metal carbonyls usually require strong nucleophiles such as organolithium reagents, and the reaction (a) produces anionic metal acyl complexes (acyl "ate" complexes) (Eq. 2.24). (For the purposes of electron bookkeeping, the resonance structure denoting both the σ-donor bond and the π–acceptor bond in metal-bound CO's is used. CO is still a formal two-electron donor ligand.) Many of these acyl "ate" complexes are quite stable and can be isolated and manipulated. Electrophiles can react at the metal (b) (analogous to C-alkylation of an enolate) giving a neutral acyl/electrophile complex, which can undergo reductive elimination (c) to produce ketones, aldehydes, or carboxylic acid derivatives. Alternatively, reaction of electrophiles at oxygen (d) (analogous to O-alkylation of an enolate) produces heteroatom-stabilized carbene complexes, which have a very rich organic chemistry in their own right (Chapter 6). The site of electrophilic attack depends both on the metal and the electrophile. Hard electrophiles such as $R_3O^+BF_4^-$, ROTf and $ROSO_2F$ undergo reaction at oxygen,[14] while softer electrophiles undergo reaction at the metal.[15]

Eq. 2.24

Ln usually a good
π-acceptor, e.g. CO

metal carbene

ketones
aldehydes
acid derivatives

Alkoxides and amines attack metal carbonyls, particularly the more reactive cationic ones (Eq. 2.25), to produce alkoxycarbonyl or carbamoyl complexes,[16] species which are involved in a variety of metal assisted carbonylation reactions via insertion into the metal-acyl carbon bond.

Eq. 2.25

alkene, alkyne
insertion

Hydroxide ion attacks metal carbonyls to produce (normally) unstable metal carboxylic acids, which usually decompose to anionic metal hydrides and carbon dioxide[17] (Eq. 2.26).

Eq. 2.26

Electrophilic carbene complexes undergo nucleophilic attack at the carbene carbon, resulting in either heteroatom exchange (Eq. 2.27)[14] or the formation of stable adducts (Eq. 2.28).[18]

Eq. 2.27

Nuc⁻ = RS⁻, R₂NH, PhLi, etc.

Eq. 2.28

stable

Nucleophilic cleavage of metal carbon σ-bonds is another process of interest in organic synthesis, since it is frequently involved in freeing an organic substrate from the metal. Many metal-acyl complexes undergo direct cleavage by alcohols or amines, a key step in many metal-catalyzed acylation reactions (Eq. 2.29).[19] Others are more stable and require an oxidative cleavage (Eq. 2.30).[20] σ-Alkyliron complexes undergo oxidative cleavage, and this process has been studied in some detail (Eq. 2.31).[21] Oxidation of the metal makes it a better leaving group, and promotes displacement with clean inversion.

Eq. 2.29

Eq. 2.30

Eq. 2.31

Another nucleophilic reaction of some consequence to organic synthesis is the self exchange reaction. Oxidative addition of organic halides to low valent metals is known to go with inversion of configuration at carbon. However, occasionally racemization or partial racemization is observed. This is thought to result from nucleophilic displacement of the metal(II) in the σ alkyl complex by the nucleophilic metal(0) starting complex (Eq. 2.32).[22] Since the reaction is "thermally neutral" this need not be a difficult process, and it should be considered whenever unexpected racemization is observed.

Eq. 2.32

Nucleophilic attack on complexed π-unsaturated hydrocarbons is among the most useful of organometallic processes for organic synthesis, since it is exactly the opposite of "normal" reaction chemistry for these substrates. Nucleophilic attack on 18 electron, cationic complexes has been particularly well-studied,[23] and, based upon a large number of examples, the following order of reactivity for kinetically controlled reactions of this class of complexes was derived:

Note that the cyclopentadienyl ligand (Cp) is among the least reactive, and, as a consequence, is often used as a spectator ligand.

The typical reaction for electrophilic metal complexes of olefins, particularly those of Pd(II), Pt(II) and Fe(II), is nucleophilic attack on the olefin.[24] In most cases, the nucleophile attacks the most substituted position of the olefin, from the face opposite the metal (*trans*), forming a carbon-nucleophile bond and a carbon metal bond (Eq. 2.33). A few nucleophiles, such as chloride, acetate, and nonstabilized carbanions, can also attack first at the metal. This is followed by insertion of the olefin, resulting in an overall *cis* addition, with attack at the less substituted position. This chemistry is normally restricted to mono and 1,2-disubstituted olefins, since more highly substituted olefins coordinate only weakly, and displacement of the olefin by the nucleophile competes.

Eq. 2.33

$Nuc^- = Cl^-, AcO^-, ROH, RNH_2, RLi$

Alkynes complexed to electrophilic metals also undergo nucleophilic attack, although this is a much less studied process, since stable alkyne complexes of electrophilic metals are rare. (These metals tend to oligomerize alkynes.) Cationic iron complexes of alkynes undergo clean nucleophilic attack from the face opposite the metal, to give stable σ-vinyl iron complexes (Eq. 2.34).[25] This process has not been used in organic synthesis to any extent.

Eq. 2.34

$Nuc^- = PhS^-, CN^-, \bar{C}H(CO_2Et)_2$ [$Ph^-, Me^-, H_2C=\bar{C}H$ $MeC\equiv C^-$ from $R_2Cu(CN)Li_2$]
$L = PPh_3, P(OPh)_3$

Nucleophilic attack on η^3-allyl metal complexes, particularly those of Pd(II), has been extensively developed as an organic synthetic method and is discussed in detail in Chapter 9. The general features are shown in Eq. 2.35. Nucleophilic attack occurs from the face opposite the metal. From a synthetic point of view, the most important reaction involving η^3-allyl complexes is the Pd(0) catalyzed reaction of allylic substrates, a process which proceeds with overall retention (two inversions) of configuration at carbon (Eq. 2.36).[26] In certain rare cases nucleophilic attack can occur at the central carbon of an η^3-allyl complex, generating a metallacyclobutane. In the case of the cationic molybdenum complex in Eq. 2.37, the complex was stable and isolated.[27]

Eq. 2.35

Eq. 2.36

Eq. 2.37

Although extended Hückel calculations[28] suggested that attack at the central carbon of η^3-allylpalladium complexes was not feasible, under appropriate conditions (a-branched ester enolates, polar solvent) just such a process occurred smoothly, to produce cyclopropanes (Eq. 2.38).[29]

Eq. 2.38

Neutral η^4-dieneiron tricarbonyl complexes undergo reaction with strong nucleophiles (e.g. $LiCMe_2CN$) at both the terminal position, to produce an η^3-allyl complex, and an internal position, to produce a σ-alkyl-η^2-olefin complex (Eq. 2.39).[30] In contrast, the much more electrophilic palladium chloride promotes exclusive terminal attack to produce stable η^3-allylpalladium complexes, which can be further functionalized (Chapter 9) (Eq. 2.40).[31] Again, nucleophilic attack occurs from the face opposite the metal.

Eq. 2.39

Eq. 2.40

In contrast to neutral η^4-diene complexes, cationic η^5-dienyl complexes are broadly reactive towards nucleophiles, and iron tricarbonyl complexes of the η^5-cyclohexadienyl ligand have found extensive use in organic synthesis. Nucleophiles ranging from electron rich aromatic compounds through organocopper species readily add to these complexes, attacking, as usual, from the face opposite the metal, to produce stable η^4-diene complexes (Eq. 2.41).[32] With substituted cyclohexadienyl ligands, the site of nucleophilic attack is generally governed by electronic factors. By coupling this nucleophilic addition to η^5-dienyl complexes with the generation of η^5-dienyl complexes by hydride abstraction from η^4-diene complexes, regio- and stereospecific polyfunctionalization of this class of ligands has been achieved (Chapter 7).

Eq. 2.41

Arenes normally undergo electrophilic attack and, except in special cases, are quite inert to nucleophilic attack. However, by complexation to electron deficient metal fragments, particularly metal carbonyls (recall that CO groups are strongly electron withdrawing), arenes become generally reactive toward nucleophiles. By far the most extensively studied complexes are the η^6-arenechromium tricarbonyl complexes (Eq. 2.42).[33] Again, the nucleophile attacks from the face opposite the metal to give a relatively unstable anionic η^5-cyclohexadienyl complex, which itself has a rich organic chemistry (Chapter 10). Oxidation regenerates the arene, while protolytic cleavage gives the cyclohexadiene.

Eq. 2.42

The reactions of nucleophiles with metal-coordinated ligands are summarized in Table 2.2 along with the d-electron count and formal oxidation states for reactants and products. Although there is a profound reorganization of electrons about the metal, most of these processes result in no change in the formal oxidation state of the metal, differentiating them from the oxidative addition/reductive elimination family of reactions. This is because in each case, a formally neutral ligand is converted to a formally mononegative ligand in the process — that is, nucleophilic attack reduces the ligand, not the metal. The two exceptions are the η^3 allyl and the η^5-dienyl complexes, in which formally negative ligands are converted to formally neutral ligands by the process and the metal is reduced.

Table 2.2   Summary of Nucleophilic Reactions on Transition Metals

Table 2.2   Continued

3.

3 neutral CO
1 neutral diene
Fe(0), d^8

2+2 e⁻

3 neutral CO
1 R(−)
overall (−)
Fe(0), d^8

4e⁻ donor

3 neutral CO

1

overall (−)
Fe(0), d^8

6e⁻ donor

4.

−CH₂CN

6e⁻ donor

3 neutral CO
1 neutral PhH
Cr(0), d^6

3 neutral CO

overall (−)
Cr(0), d^6

BUT

4e⁻ donor

Formally a
reduction

1 Cl⁻, 1
1 neutral PPh₃
overall neutral
Pd²⁺, d^8

2e⁻ donor
other 2e⁻
go to reduce
metal

1 neutral C=C
1 neutral
PPh₃
overall neutral
Pd(0), d^{10}

2.6 Transmetallation

Transmetallation is a process of increasing importance in the area of transition metals in organic synthesis, but has been little studied and is not well understood. The general phenomenon involves the transfer of an R group from a main group organometallic compound to a transition metal complex (Eq. 2.43).[34] When combined with reactions which introduce an R group into the transition metal complex such as oxidative addition or nucleophilic attack on alkenes, efficient carbon-carbon bond forming (cross coupling) reactions ensue. In cases such as these, the transmetallation step is almost always the rate limiting step and when catalytic cycles involving a transmetallation step fail, it is usually this step which needs attention.

Eq. 2.43

$$RM \ + \ M'X \ \underset{}{\overset{K}{\rightleftharpoons}} \ RM' \ + \ MX$$

M = Zn, Zr, B, Hg, Si, Sn, Ge
M' = transition metal

In order for the transmetallation step to proceed, the main group organometallic M must be more electropositive than the transition metal M'. However, since this is an equilibrium, if RM' is irreversibly consumed in a subsequent step, the process can be used even if it is not favorable (K is small). This, coupled with uncertain accuracy of numbers for electronegativity, make experiment the wisest course.

An often neglected fact is that Eq. 2.43 *is* an equilibrium, and that both partners must profit, thermodynamically, from the process. Therefore, the nature of X may well be as important as the nature of M or M', and transmetallation procedures can often be promoted by adding appropriate counterions, X (Chapter 4).

2.7 Electrophilic Attack on Transition Metal Coordinated Ligands

Reactions of coordinated organic ligands with electrophiles has found substantially less use in organic synthesis than have reactions with nucleophiles. Perhaps the most common electrophilic reaction is the electrophilic cleavage of metal-carbon sigma bonds, as a method to free an organic substrate from a metal template. If the metal complex has d electrons, electrophilic attack usually occurs at the metal, a formal oxidative addition. This is followed by reductive elimination to result in cleavage with retention of configuration at carbon (Eq. 2.44). If a sufficiently good nucleophile is present (e.g. Br⁻ from Br_2), cleavage with *inversion* may result. An example of this kind of cleavage has already been discussed (Eq. 2.31). Even σ-alkyl complexes of metals having no d electrons undergo facile electrophilic cleavage, and usually with retention of configuration. In these cases, direct electrophilic attack at the metal-carbon σ-bond (S_E2) is the likely process (Eq. 2.45).[35]

Eq. 2.44

Eq. 2.45

Electrophiles also can react at *carbon* in organometallic complexes, and a variety of useful processes involve this type of reaction. These are summarized in Table 2.3. Some, particularly reactions of electrophiles with η^1-allyl complexes (γ attack) and hydride abstraction from η^4-diene complexes, are quite useful in organic synthesis and are discussed in detail later.

Table 2.3 Electrophilic Attack on Transition Metal Coordinated Organic Ligands

1. Electrophilic cleavage of σ-alkyl metal bonds.

$$R{-}M \ + \ E^+ \longrightarrow \ R{-}E \qquad \underline{retention} \ at \ R$$

2. Attack at α-position

3. Attack at β-position

4. Attack at γ-position

Table 2.3 Continued

5. Attack on coordinated polyenes

References

1. (a) Collman, J.P.; Hegedus, L.S.; Norton, J.R.; Finke, R.G. *Principles and Applications of Organotransition Metal Chemistry*, 2nd ed. University Science Books, Mill Valley, CA, 1987, pp. 235-458. (b) Twigg, M.V., ed. *Mechanisms of Inorganic and Organometallic Reactions*. Plenum, New York, annually since 1983, see part 3 in each volume.
2. (a) Odell, L.A.; Raethel, H.A. *Chem. Comm.* **1968**, 1323; **1969**, 87.
3. Tolman, C.A. *Chem. Rev.* **1977**, *77*, 313.
4. Albers, M.O.; Coville, N.J. *Coord. Chem. Rev.* **1984**, *53*, 227.
5. (a) Darensbourg, D.J.; Walker, N.; Darensbourg, M. *J. Am. Chem. Soc.* **1980**, *102*, 1213. (b) Darensbourg, D.J.; Darensbourg, M.Y.; Walker, N. *Inorg. Chem.* **1981**, *20*, 1918. (c) Darensbourg, D.J.; Ewen, J.A. *Inorg. Chem.* **1981**, *20*, 4168.
6. For reviews see: (a) Collman, J.P.; Roper, W.R. *Adv. Organomet. Chem.* **1968**, *7*, 53. (b) Collman, J.P. *Acc. Chem. Res.* **1968**, *1*, 136. (c) Stille, J.R.; Lau, K.S.Y. *Acc. Chem. Res.* **1977**, *10*, 434. (d) Stille, J.K. in *The Chemistry of the Metal-Carbon s-Bond*, vol. 2, Hartley, F.R.; Patai, S., eds., Wiley, New York, 1985, pp. 625-787.
7. (a) Sweany, R.L. *J. Am. Chem. Soc.* **1985**, *107*, 2374. (b) Upmacis, R.K.; Gadd, G.E.; Poliakoff, M.; Simpson, M.B.; Turner, J.J.; Whyman, R.; Simpson, A.F. *J. Chem. Soc. Chem. Comm.* **1985**, 27.
8. Chock, P.B.; Halpern, J. *J. Am. Chem. Soc.* **1966**, *88*, 3511.
9. Collman, J.P.; Finke, R.G.; Cawse, J.N.; Brauman, J.I. *J. Am. Chem. Soc.* **1977**, *99*, 2515.
10. (a) Labinger, J.R.; Osborn, J.A. *Inorg. Chem.* **1980**, *19*, 3230. (b) Labinger, J.A.; Osborn, J.A.; Coville, N.J. *Inorg. Chem.* **1980**, *19*, 3236. (c) Jenson, F.R.; Knickel, B. *J. Am. Chem. Soc.* **1971**, *93*, 6339.
11. Kochi, J.K. *Pure Appl. Chem.* **1980**, *52*, 571.
12. Gillie, A.; Stille, J.K. *J. Am. Chem. Soc.* **1980**, *102*, 4933.
13. For a review see: Hegedus, L.S. "Nucleophilic Attack on Transition Metal Organometallic Compounds" in *The Chemistry of the Metal-Carbon s-Bond*, vol. 2, Hartley, F.R.; Patai, S., eds., Wiley, New York, 1985, pp. 401-512.
14. Dötz, K.H. *Angew. Chem. Int. Ed. Engl.* **1984**, *23*, 587.
15. Collman, J.P. *Acc. Chem. Res.* **1975**, *8*, 342.
16. Angelici, R.L. *Acc. Chem. Res.* **1972**, *5*, 335.
17. Ungermann, C.; Landis, V.; Moya, S.A.; Cohen, H.; Walker, H.; Pearson, R.G.; Rinker, R.G.; Ford, P.C. *J. Am. Chem. Soc.* **1979**, *101*, 5922 and references therein.
18. Casey, C.P.; Miles, W.M. *J. Organomet. Chem.* **1983**, *254*, 333.
19. Heck, R.F. *Pure Appl. Chem.* **1978**, *50*, 691.

20. Hegedus, L.S.; Anderson, O.P.; Zetterberg, K.; Allen, G.; Siirala-Hansen, K.; Olsen, D.J.; Packard, A.B. *Inorg. Chem.* **1977**, *16*, 1887.
21. Whitesides, G.M.; Boschetto, D.J. *J. Am. Chem. Soc.* **1971**, *93*, 1529.
22. Lau, K.S.Y..; Fries, R.M.; Stille, J.K. *J. Am. Chem. Soc.* **1974**, *96*, 4983.
23. Davies, S.G.; Green, M.L.H.; Mingos, D.M.P. *Tetrahedron* **1978**, *34*, 3047.
24. (a) Lennon, P.M.; Rosan, A.M.; Rosenblum, M. *J. Am. Chem. Soc.* **1977**, *99*, 8426. (b) Hegedus, L.S.; Akermark, B.; Zetterberg, K.; Olsson, L.F. *J. Am. Chem. Soc.* **1984**, *106*, 7122. (c) Hegedus, L.S.; Williams, R.E.; McGuire, M.A.; Hayashi, T. *J. Am. Chem. Soc.* **1980**, *102*, 4973.
25. Reger, D.L.; Belmore, K.A.; Mintz, E.; McElligott, P.J. *Organometallics* **1984**, *3*, 134; 1759.
26. Hayashi, T.; Hagihara, T.; Konishi, M.; Kumada, M. *J. Am. Chem. Soc.* **1983**, *105*, 7767.
27. Periana, R.A.; Bergman, R.G. *J. Am. Chem. Soc.* **1984** *106*, 7272.
28. Curtis, M.D.; Eisenstein, O. *Organometallics* **1984**, *3*, 887.
29. (a) Hegedus, L.S.; Darlington, W.H.; Russel, C.E. *J. Org. Chem.* **1980**, *45*, 5193. (b) Carfagna, C.; Mariani, L.; Musco, M.; Sallese, G.; Santi, R. *J. Org. Chem.* **1991**, *56*, 3924.
30. (a) Semmelhack, M.F.; Le, H.T.M. *J. Am. Chem. Soc.* **1984**, *106*, 2715. (b) Semmelhack, M.F.; Herndon, J.W. *Organometallics* **1983**, *2*, 363.
31. Bäckvall, Jan-E.; Nyström, Jan-E.; Nordberg, R.E. *J. Am. Chem. Soc.* **1985**, *107*, 3676.
32. Pearson, A.J. *Comp. Org. Synth.* **1991**, *4*, 663.
33. Semmelhack, M.F. *Comp. Org. Synth.* **1991**, *4*, 517.
34. Negishi, E-I. *Organometallics in Organic Synthesis*. Wiley, New York, 1980.
35. Labinger, J.A.; Hart, D.W.; Seibert, W.E., III; Schwartz, J. *J. Am. Chem. Soc.* **1975**, *97*, 3851.

CHAPTER 3

Synthetic Applications of Transition Metal Hydrides

3.1 Introduction

Transition metal hydrides are an important class of organometallic complex, primarily because of their role in homogeneous hydrogenation, hydroformylation, and hydrometallation reactions. The reactivity of transition metal hydrides depends very much on the metal and the other ligands, and ranges all the way from hydride donors to strong protic acids![1] However, their major use in synthesis relies on neither of these characteristics, but rather their propensity to insert alkenes and alkynes into the M-H bond, generating σ-alkylmetal complexes for further transformations.

Transition metal hydrides can be synthesized in a number of different ways (Figure 3.1), but the most commonly used method is oxidative addition. Transition metals are unique in their ability to activate hydrogen under very mild conditions, a feature which accounts for their utility in catalytic hydrogenation.

Figure 3.1 Syntheses of Transition Metal Hydrides

3.2 Homogeneous Hydrogenation

When synthetic chemists are faced with the reduction of a carbon-carbon double bond, their first choice is, almost invariably, *heterogeneous* catalysis, using something like palladium on carbon. Heterogeneous catalysts are convenient to handle, efficient, and easy to remove by filtration after the reaction is complete. In contrast, homogeneous catalysts are sometimes difficult to handle, sensitive to impurities, such as traces of oxygen, have a tendency to cause olefin isomerization, and are difficult to recover, because they are soluble. However, their enormous advantage, selectivity, mitigates these shortcomings.

There are two general classes of homogeneous hydrogenation catalysts — monohydrides and dihydrides — and they react by different mechanisms and have different specificities. The best-studied monohydride catalyst is the lemon yellow, air stable, crystalline solid $Rh(H)(PPh_3)_3(CO)$, and the mechanism by which it hydrogenates olefins is shown in Figure 3.2. It is *completely* specific for terminal olefins, and will tolerate internal olefins, aldehydes, nitriles, esters, and chlorides elsewhere in the molecule. Its major limitation is that olefin isomerization competes with hydrogenation, and the isomerized olefin cannot be reduced by the catalyst.

The starting complex is a coordinatively saturated, Rh(I) d^8 complex and the first step requires loss of a phosphine (ligand dissociation). As a consequence, added phosphine inhibits catalysis. Coordination of the substrate alkene to the unsaturated metal center is followed by migratory insertion, to produce an unsaturated σ-alkylrhodium(I). The regiochemistry of insertion is not known, and the rhodium may occupy either the terminal or internal position. Oxidative addition of hydrogen to this σ-alkylrhodium(I) complex, followed by reductive elimination of R–H, results in reduction of the alkene and regeneration of the catalytically active, unsaturated rhodium(I) monohydride. Irreversible olefin isomerization occurs if a 2° σ-alkylrhodium(I) intermediate undergoes β-hydrogen elimination (the reverse of migratory insertion) *into* the alkyl chain, producing an internal olefin and the catalytically active rhodium(I) monohydride. Because this catalyst cannot reduce internal olefins, this isomerization is irreversible, and competitive.

Dihydride catalysts are much more versatile, and much better understood. The most widely used catalyst precursor is the burgundy red

Figure 3.2 Reduction of Alkenes by Monohydride Catalysts

Catalytic Cycle

Figure 3.3 Reduction of Alkenes by Wilkinson's Catalyst

solid $RhCl(PPh_3)_3$, Wilkinson's complex. The mechanism of hydrogenation of alkenes by this catalyst has been studied in great detail, and is quite complex (Figure 3.3).[2] Although Wilkinson's compound is already coordinatively unsaturated (Rh^I, d^8, 16e−) the kinetically active species is the 14 electron complex $RhCl(PPh_3)_2$, with, perhaps, solvent loosely associated. In this case, oxidative addition of H_2 is the initial step, followed by coordination of the alkene. Migratory insertion is the rate limiting step, followed by a fast reductive elimination of the alkane and regeneration of the active catalyst. In addition to the catalytic cycle, several other equilibria are going on, complicating the system.

Wilkinson's catalyst is attractive for synthesis because it is selective and efficient, tolerant of functionality, and, in contrast to monohydride catalysts, does not promote olefin isomerization. The reactivity towards olefins parallels their coordination ability:

with a 50-fold difference in rate over this range of alkenes. Noncyclic tri- and tetrasubstituted olefins are not reduced, and ethylene, which coordinates too strongly, poisons the catalyst. Alkynes are rapidly reduced. Nitro olefins are reduced to nitro alkanes without competitive reduction of the nitro goup.

There are several limitations of reductions using Wilkinson's catalyst. In solution, it catalyzes the oxidation of triphenylphosphine, which in turn, leads to destruction of the complex, so air cannot be tolerated. Strong ligands such as ethylene, thiols, and very basic phosphines poison the system. Carbon monoxide, and substrates prone to decarbonylation, such as acid halides, are not tolerated. Notwithstanding these limitations, Wilkinson's catalyst is quite useful in synthesis, as exemplified by Equations 3.1 and 3.2.[3]

Eq. 3.1

Eq. 3.2

Of more interest to synthetic chemists is asymmetric homogeneous hydrogenation, wherein prochiral olefins are reduced to enantiomerically enriched products. This topic has been studied with excruciating detail, and when it works, it is spectacular! However, the requirements for successful asymmetric hydrogenation are so stringent that, until very recently, the process was limited to a very narrow range of substrates.

The catalysts used are cationic rhodium(I) complexes of optically active chelating diphosphines, such as DIOP, DIPAMP, CHIRAPHOS and BPPM (Figure 3.4), generated *in situ* by reduction of the corresponding diphosdiolefin complex (Eq. 3.3). These catalysts are extremely efficient in the asymmetric hydrogenation of prochiral olefins which can bind in a

Eq. 3.3

S = Solvent

bidentate, chelate manner, such as α-acetamidoacrylates, which are reduced to optically active α-amino acids. This system has been studied in minute detail, and several key intermediates in the proposed catalytic cycle have been fully characterized by ^{1}H, ^{13}C, ^{31}P NMR spectroscopy and by X-ray crystallography. All indicate that the substrate olefin coordinates in a bidentate manner, with the amide oxygen acting as the second ligand, and that two diastereoisomeric olefin complexes can be formed. The only problem with all of these studies was that the diastereoisomer detected would lead to the opposite absolute configuration of the product than was observed! Careful kinetic studies[4] resolved this problem, and led to the following (after the fact) obvious observation that, if you can detect intermediates in a catalytic process, they probably are not involved. That is, anything that accumulates sufficiently to be detected is kinetically inactive; kinetically active species just react, converting starting material to product, and are never present in any appreciable amount.

The mechanism for asymmetric hydrogenation of acetamidoacrylates is shown in Figure 3.5. Its importance to synthetic chemists lies not with its details, but rather the clear illustration it provides of how closely balanced a series of complex steps must be to achieve efficient asymmetric hydrogenation. It is for this reason that the range of prochiral olefins asymmetrically hydrogenated by these catalysts is so narrow, and why extrapolation to different substrates is difficult.

Figure 3.4 Optically Active Phosphines

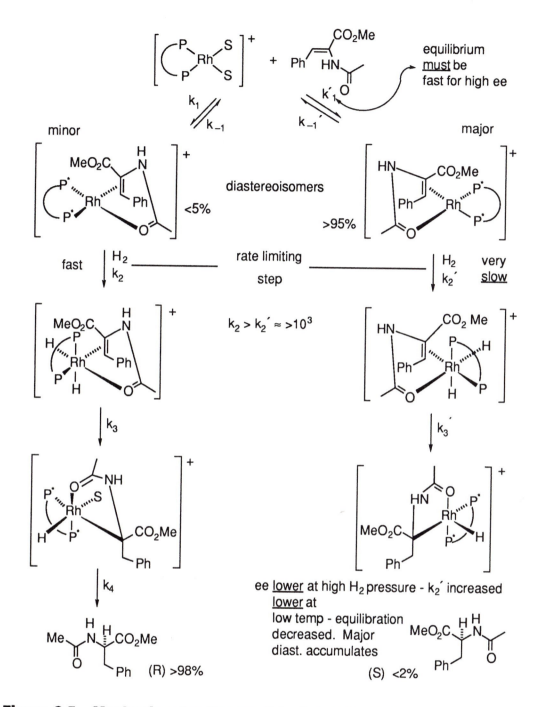

Figure 3.5   Mechanism for Asymmetric Catalytic Hydrogenation of Acetamidoacrylates

RuX$_2$ (BINAP)   X = Br$^-$, Cl$^-$, OAc$^-$, CF$_3$CO$_2^-$

BINAP =

Figure 3.6   Ruthenium BINAP Catalyst

There are two features of the process critical to highly asymmetric induction; the two diastereoisomers must undergo reaction at substantially different rates, and they must equilibrate rapidly. The rate limiting step is the oxidative addition of hydrogen to the rhodium(I) olefin complex, and one (the minor) diastereomer undergoes this reaction at a rate 10^3 greater than the other. Provided the two diastereoisomeric olefin complexes can equilibrate rapidly, highly asymmetric induction can be observed. Interestingly, higher hydrogen pressure decreases the enantioselectivity, since the oxidative addition rate for both diastereoisomers is increased.

A much more broadly useful class of asymmetric hydrogenation catalysts, which also relies upon bidentate chelate coordination of the olefinic substrate, is the ruthenium(II) BINAP system, where BINAP is an axially dissymmetric diphosphine (Figure 3.6).[5] These complexes reduce a wide range of substrates with high asymmetric induction, relying on the rigidity of complexation confered by chelation to assure high enantiose were reduced selectively and efficiently (Eq. 3.4),[6] and a very wide range of functionality was tolerated. The antiinflammatory Naproxen was produced with 97% ee in a high-pressure reduction in methanol, as was a thienamycin precursor, albeit with lower enantioselectivity.

N-Acylenamines were also efficient coordinating groups and 1,2,3,4-tetrahydroisoquinoline derivatives were produced in 95-100% ee by reduction under 1-4 atm of hydrogen in methanol (Eq. 3.5).[7]

Eq. 3.4

also

97% ee (Naproxen)

74% de (Thienamycin)

R^1	R^2	R^3	ee
Me	Me	H	91
H		Me	87
H	Me	Ph	85
Ph	H	H	92
H	$HOCH_2$	Me	93
H	CH_3	$COOCH_2CMe$	95

Eq. 3.5

95-100% ee

Allylic and homoallylic alcohols also undergo selective reduction with high enantioselectivity (Eq. 3.6).[8] Geraniol was hydrogenated to citronellol with high ee, with no net reduction of the remote double bond (Eq. 3.7). Either geometrical isomer could be converted to either enantiomer by appropriate choice of BINAP ligand. Homogeraniol was also reduced selectively, but *bis*-homogeraniol was inert. This indicates that bidentate, chelate coordination of the olefinic substrates is required for *reactivity* as well as enantioselectivity. This feature was used to synthesize dolichols, wherein only the allylic alcohol olefin was reduced (Eq. 3.8).[9]

Eq. 3.6

Eq. 3.7

NOTE

(Homo) 92% ee

∴ OH directs

but inert

Eq. 3.8

95% ee <2%
reduction of other
C=C

Racemic secondary allylic alcohols could be efficiently resolved by carrying out a partial hydrogenation over a ruthenium BINAP catalyst (Eq. 3.9).[10] A wide variety of allylic alcohols undergo this reaction efficiently.

Eq. 3.9

Most homogeneous hydrogenation catalysts are specific for carbon-carbon double bonds, and other multiply-bonded species are inert. In marked contrast, ruthenium-BINAP complexes catalyze the efficient asymmetric reduction of the carbonyl group, provided there is a heteroatom in the α, β, or γ position to provide the requisite second point of attachment to the catalyst (Eq. 3.10). β-Keto ester compounds were efficiently reduced to β-hydroxy esters by this system (Eq. 3.11).[11] Racemic β-keto esters were reduced with high selectivity for each diastereoisomer (Eq. 3.12). However, when the reaction was carried out under conditions which ensured rapid equilibration of the α-position, high yields of a single diastereoisomer were obtained (Eq. 3.13).[12]

Eq. 3.10

n = 1-3, X = heteroatom, C = sp^2, sp^3

Eq. 3.11

R = Me, Et, nPr, t-Bu, Ph
R^1 = Me, Et, iPr, t-Bu

>99% yield
99% ee

Eq. 3.12

97% ee   49%   96% ee   51%

Eq. 3.13

>95:5 syn/anti
92-98% ee

R^1 = Me,   R^2 = NHAc, NHCO$_2$Bn, CH$_2$NHAc   cat. = (R) RuBINAP

93-99% ee

Figure 3.6 Substrates Efficiently Reduced Asymmetrically by RuBINAP Catalysts

This asymmetric reduction is not restricted to β-keto esters, but rather proceeds efficiently with a wide range of carbonyl compounds having appropriately disposed directing groups (Figure 3.6).[13]

Another synthetically useful class of hydrogenation catalyst is the "super unsaturated" iridium or rhodium complex catalyst system, known as Crabtree's catalysts.[14] These are generated *in situ* by reduction of cationic cyclooctadienyl iridium(I) complexes in the noncoordinating solvent, dichloromethane (Eq. 3.14). The solvent is apparently so loosely coordinated that it is replaced by virtually any olefinic substrate. These catalysts are exceptionally efficient for the reduction of tetrasubstituted olefins, substrates most hydrogenation catalysts will not reduce. A comparison of Crabtree's catalyst with Wilkinson's catalyst is shown in Figure 3.7. Nonpolar solvents must be used with Crabtree's catalysts. Acetone and ethanol act as competitive inhibitors, and greatly reduce catalytic activity.

Eq. 3.14

(Cy = cyclohexyl)

| "L$_2$Ir" | 6400 | 4500 | 4000 |

vs

| L$_3$RhCl | 650 | 700 | 0 |

Figure 3.7   Comparison of Wilkinson's and Crabtree's Catalysts

The observation that Crabtree's catalyst coordinates to alcohols has been used to an advantage in synthesis, by using adjacent "hard" ligands in the substrate to coordinate to the catalyst and deliver it from a single face of the alkene. Equation 3.15 compares Crabtree's catalyst to the normally used palladium on carbon. The iridium catalyst was delivered exclusively from the same face of the olefin as that occupied by the OH group, while Pd/C was nonselective.[15] This same effect was seen with a much more complex substrate (Eq. 3.16).[16] In both cases, the rate of hydrogenation was slower than that for substrate without coordinating functional groups.

Eq. 3.15

| Pd/C | 20 | : | 30 |
| L$_2$Ir | 99.9 | : | <0.1 |

Eq. 3.16

(10 cases studied)

| L$_2$Ir$^+$ | >99:1 |
| Pd/C | 1:2 |

Another catalytic hydrogenation that involves a somewhat different form of "two point" attachment is the 1,4-reduction of 1,3-dienes to Z-alkenes in the presence of group 6 catalyst precursors such as $L_3Cr(CO)_3$ where L_3 is an arene,[17] $(CH_3CN)_3$ or $(CO)_3$[18] (Eq. 3.17). This reduction is restricted to dienes which can achieve cisoid conformations, implying the intermediacy of a chelate diene complex in the process. These same complexes catalyze the reduction of α,β-unsaturated carbonyl compounds, but again only those which can achieve an S-cis conformation (Eq. 3.18).[19] In the absence of hydrogen, these complexes catalyze the rearrangement of p(1Z)-1-[silyloxymethyl]butadiene to silyl dienol ethers with high stereoselectivity (Eq. 3.19).[20]

Eq. 3.17

Eq. 3.18

and

Eq. 3.19

3.3 Other Reductions

In all of the reactions presented above, molecular hydrogen was the source of hydrogen for the reduction, and the ability of transition metals to break the hydrogen-hydrogen bond in an oxidative addition reaction was central to the process. However, metal hydrides are produced by several other pathways, and some of these have proven synthetically useful.

Treatment of iron pentacarbonyl with KOH produces an iron hydride which efficiently reduces α,β-unsaturated aldehydes, ketones, esters, lactones, and nitriles to saturated derivatives, with no reduction of the carbonyl or nitrile group (Eq. 3.20).[21] The reaction proceeds by rapid, irreversible addition of H–Fe across the double bond, followed by protonolysis.[22] Thus, in this reduction, both hydrogens came from water. The related binuclear complex $NaHFe_2(CO)_8$ affects similar chemistry but the reaction proceeds by a somewhat different mechanism.[22] Both systems are inefficient with β-substituted or sterically hindered enones.

Eq. 3.20

The hexamer $[(Ph_3P)CuH]_6$ reduces conjugated enones to saturated ketones stoichiometrically, presumably via a conjugate addition to produce a copper enolate intermediate.[23,24] Interestingly, under modest hydrogen pressure in the presence of excess phosphine, this hexamer catalyzed this reduction.[24]

Yet another efficient system for the reduction of conjugated enones is the combination of a palladium(0) catalyst and tributyltin hydride (Eq. 3.21).[25] The mechanism of the process is not known. A possibility is shown in Eq. 3.22.

Eq. 3.21

(lactone not reduced)

Eq. 3.22

3.4 Other Reactions of Transition Metal Hydrides

Transition metal hydrides are involved in a host of other synthetically useful reactions, most of which proceed by "hydrometallation" (insertion into a metal-hydrogen bond) of an alkene or alkyne to produce σ-alkylmetal complexes. It is the subsequent reactions of these σ-alkylmetal species that provide the synthetically useful transformations, and these will be discussed in the next chapter.

References

1. Moore, E.J.; Sullivan, J.M.; Norton, J.R. *J. Am. Chem. Soc.* **1986**, *108*, 2257.
2. Halpern, J.; Okamoto, T.; Zakhariev, A. *J. Mol. Catal.* **1976**, 2, 65.
3. Koch, G.K.; Dalenberg, J.W. *J. Labelled Compounds* **1970**, 6, 395.
4. Halpern, J. *Science* **1982**, 217, 401.
5. Noyori, R.; Takaya, H. *Acct. Chem. Res.* **1990**, 23, 345.
6. Ohta, T.; Takaya, H.; Kitamura, M.; Nagai, K.; Noyori, R. *J. Org. Chem.* **1987**, 52, 3174.
7. Noyori, R.; Ohta, M.; Hsiao, Y.; Kitamura, M. *J. Am. Chem. Soc.* **1986**, *108*, 7117.
8. Takaya, H.; Ohta, T.; Sayo, N.; Kumobayashi, H.; Akutagawa, S.; Inoue, S-I.; Kasahara, I.; Noyori, R. *J. Am. Chem. Soc.* **1987**, *109*, 1596.
9. Imperiali, B.; Zimmerman, J.W. *Tetrahedron Lett.* **1988**, 29, 5343.
10. Kitamura, M.; Kasahara, I.; Manabe, K.; Noyori, R.; Takaya, H. *J. Org. Chem.* **1988**, 53, 708.
11. Noyori, R.; Ohnkuma, T.; Kitamura, M.; Takaya, H.; Sayo, N.; Kamobayashi, H.; Akutagawa, S. *J. Am. Chem. Soc.* **1987**, *109*, 5856.
12. Noyori, R.; Ikeda, T.; Ohkuma, T.; Widhalm, M.; Kitamura, M.; Takaya, H.; Akutagawa, S.; Sayo, N.; Saito, T.; Taketomi, T.; Kumobayashi, H. *J. Am. Chem. Soc.* **1989**, *111*, 9134.
13. Kitamura, M.; Ohnkuma, T.; Inoue, S.; Sayo, N.; Kumobayashi, H.; Akutagawa, S.; Ohta, T.; Takaya, H.; Noyori, R. *J. Am. Chem. Soc.* **1988**, *110*, 629.
14. Crabtree, R. *Acct. Chem. Res.* **1979**, 12, 331.
15. Crabtree, R.H.; Davis, M.W. *Organometallics* **1983**, 2, 681.
16. Schultz, A.G.; McCloskey, P.J. *J. Org. Chem.* **1985**, 50, 5905.
17. (a) Tucker, J.R.; Riley, D.P. *J. Organomet. Chem.* **1985**, 279, 49. (b) LeMaux, P.; Jaouen, G.; Saillard, J-Y. *J. Organomet. Chem.* **1981**, 212, 193.
18. (a) Wrighton, M.S.; Schroeder, M.A. *J. Am. Chem. Soc.* **1973**, 95, 5764. (b) Mirbach, M.J.; Tuyet, N.P.; Saus, A. *J. Organomet. Chem.* **1982**, 236, 309.
19. Sodeoka, M.; Shibasaki, M. *J. Org. Chem.* **1985**, 50, 1147; for a review see Sodeoka, M.; Shibasaki, M. *Synthesis* **1993**, 643.
20. Sodeoka, M.; Yamada, H.; Shibasaki, M. *J. Am. Chem. Soc.* **1990**, 112, 4907.
21. Noyori, R.; Umeda, I.; Ishigami, T. *J. Org. Chem.* **1972**, 37, 1542.
22. Collman, J.P.; Finke, R.G.; Matlock, P.L.; Wahren, R.; Komoto, R.G.; Brauman, J.I. *J. Am. Chem. Soc.* **1978**, *100*, 1119.

23. Mahoney, W.S.; Brestensky, D.M.; Stryker, J.M. *J. Am. Chem. Soc.* **1988**, *110*, 291.
24. Mahoney, W.S.; Stryker, J.M. *J. Am. Chem. Soc.* **1989**, *111*, 8818.
25. Keinan, E.; Gleize, P.A. *Tetrahedron Lett.* **1982**, *23*, 477.

CHAPTER 4

Synthetic Applications of Complexes Containing Metal-Carbon σ-Bonds

4.1 Introduction

Transition metal complexes containing metal-carbon σ-bonds are central to a majority of transformations in which transition metals are used to form carbon-carbon and carbon-X bonds, and thus are of supreme importance for organic synthesis. The transition metal-to-carbon σ-bond is usually covalent rather than ionic, and this feature strongly moderates the reactivity of the bound organic group, restricting its reactions to those accessible to the transition metal (e.g., oxidative addition, insertion, reductive elimination, β-hydrogen elimination). That is, the transition metal is much more than a sophisticated counter ion for the organic group; it is the major determinant of the chemical behavior of that organic group.

σ-Carbon-metal complexes can be prepared by a number of methods, summarized in Figure 4.1. This variety makes virtually every class of organic compound a potential source of the organic group in these complexes, and thus subject to all of the carbon-carbon bond forming reactions of σ-carbon-metal complexes. The use of these processes in organic synthesis is the subject of this chapter.

1. Preparation

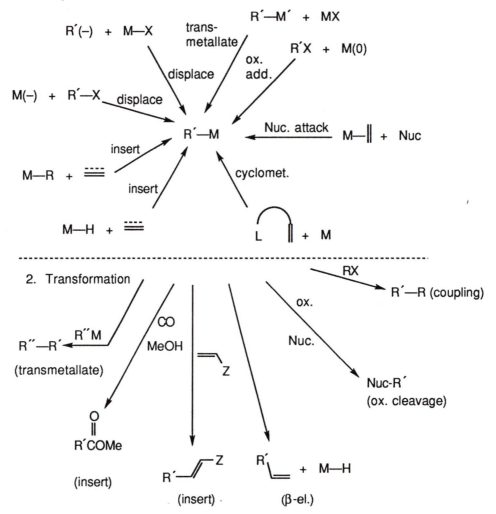

Figure 4.1 **Preparation and Transformation of σ-Carbon-Metal Complexes**

4.2 σ-Carbon-Metal Complexes from the Reaction of Carbanions and Metal Halides: Organocopper Chemistry

Starting with the early observation that small amounts of copper(I) salts catalyzed the 1,4-addition of Grignard reagents to conjugated enones,[1] organocopper chemistry has been warmly accepted by synthetic chemists and vastly widened in scope by organometallic chemists. As a broad class, they are among the most extensively used organometallic reagents, and the variety of species and the transformations they affect is staggering.[2,3]

The simplest and most extensively used complexes are the lithium diorganocuprates, R_2CuLi, soluble, thermally unstable complexes generated *in situ* by the reaction of copper(I) iodide with two equivalents of an organo-lithium reagent (Eq. 4.1). (The *mono* alkyl copper complexes, RCu, are yellow, insoluble, oligomeric complexes, little used in synthesis.) These reagents efficiently alkylate a variety of halides (Eq. 4.2).

Eq. 4.1

$$2RLi + CuI \longrightarrow R_2CuLi + LiI$$

Eq. 4.2

$$R_2CuLi + R^1X \longrightarrow R-R^1$$

The range of useful halide substrates is very broad. The order of reactivity of alkyl (sp^3) halides is primary > secondary >> tertiary, with iodides more reactive than bromides and chlorides. Alkyl tosylates also undergo substitution by diorganocuprates. Alkenyl halides are also very reactive toward lithium diorganocuprates, and react with essentially complete retention of geometry of the double bond, a feature which has proved useful in organic synthesis. Aryl iodides and bromides are alkylated by lithium organocuprates, although halogen-metal exchange is sometimes a problem. (RCu reagents are sometimes more efficient in this reaction.) Benzylic, allylic, and propargylic halides also react cleanly. Propargyl halides produce primarily allenes (via S_N2'-type reactions), but allylic chlorides and bromides react without allylic transposition. In contrast, allylic acetates react mainly with allylic transposition. Acid halides are converted to ketones by lithium dialkyl-cuprates, although alkyl hetero (mixed) cuprates are more efficient in this process. Finally, epoxides undergo ring opening resulting from alkylation at the less-substituted carbon. Consideration of a great deal of experimental data leads to the following order of reactivity of lithium diorganocuprates with organic substrates: acid chlorides > aldehydes > tosylates ≈ epoxides > iodides > ketones > esters > nitriles. This is only a generalization; many exceptions are caused by unusual features in either the organocopper reagent or the substrate.

Another extensively used process in the synthesis of complex organic molecules is the conjugate addition of lithium diorganocuprates to enones (Eq. 4.3). The range of R groups available for this useful reaction is similar to that for the alkylation reaction. Virtually all types of sp^3 hybridized dialkyl-cuprate reagents react cleanly, including straight-chain and branched primary, secondary, and tertiary alkyl complexes. Phenyl and substituted aryl-cuprates react in a similar fashion, as do vinylcuprates. Allylic and benzyl-cuprates also add 1,4 to conjugated enones, but they are no more efficient than the corresponding Grignard reagents. Acetylenic copper reagents do *not* transfer an alkyne group to conjugated enones, a feature used to advantage in the chemistry of mixed cuprates.

Eq. 4.3

$$R_2CuLi \; + \; CH_2{=}CHCCH_3 \;\longrightarrow\; RCH_2CH_2CCH_3$$

The range of α,β-unsaturated carbonyl substrates that undergo 1,4-addition with diorganocuprates is indeed broad. Although the reactivity of any particular system depends on many factors, including the structure of the copper complex and the substrate, sufficient data are available to allow some generalizations. Conjugated ketones are among the most reactive substrates, combining very rapidly with diorganocuprates (the reaction is complete in less than 0.1 s at 25°) to produce the 1,4 adduct in excellent yield. Alkyl substitution at the α, α', or β positions of the parent enone causes only slight alterations in the rate and course of this conjugate addition with acyclic enones, although this same substitution, as well as remote substitution, does have stereochemical consequences with cyclic enone systems.

Conjugated esters are somewhat less reactive than conjugated ketones. With such esters, α,β- and β,β-disubstitution decreases reactivity drastically. Conjugated carboxylic acids do not react with lithium diorganocuprates, and conjugated aldehydes suffer competitive 1,2-additions. Conjugated anhydrides and amides have not been studied to any extent.

Conjugate additions of organocuprates to enones initially produces an enolate, and further reaction with an electrophile is possible, although often times slow and inefficient. This problem has been overcome, in the context of prostaglandin synthesis, by use of a butylphosphine stabilized copper reagent, and by conversion of the unreactive, initially formed copper (or lithium) enolate into a more reactive tin enolate by addition of triphenyl tin chloride (Eq. 4.4).[4]

Eq. 4.4

76%

Although lithium dialkylcuprates are very useful reagents, they suffer several serious limitations, the major one being that only one of the two alkyl groups is transferred. Because of the instability of the reagent, a threefold to fivefold excess (a sixfold to tenfold excess of R) is often required for complete reaction. When the R group is large or complex, as it often is in prostaglandin chemistry, this lack of efficiency is intolerable. Cuprous alkoxides, mercaptides, cyanides, and acetylides are considerably more stable than cuprous alkyls, and, furthermore, are relatively unreactive. By making mixed alkyl heterocuprates with one of these nontransferable, stabilizing groups, this problem can be avoided.[5] Among the heterocuprates, lithium phenylthio-(alkyl)cuprates and lithium *t*-butoxy(alkyl)cuprates are the most useful, and are stable at about –20° to 0°C.

Mixed alkyl heterocuprates are the most efficient reagents for the conversion of acid chlorides to ketones. Although large excesses of lithium dialkylcuprates are required for this reaction, and secondary and tertiary organocuprates never work well, only a 10 percent excess of these mixed cuprates is required to provide high yields of ketones, even in the presence of remote halogen, keto, and ester groups. In addition, although aldehydes are unstable toward lithium dialkylcuprates even at –90°, equal portions of benzaldehyde and benzoyl chloride react with one equivalent of lithium phenylthio(t-butyl)cuprate to produce pivalophenone in 90% yield with 73% recovery of benzaldehyde. The lithium phenylthio(t-alkyl)cuprate is also the reagent of choice for the alkylation of primary alkyl halides with t-alkyl groups. Secondary and tertiary halides fail to react with this reagent, however.

Perhaps most importantly, these mixed cuprate reagents, especially the acetylide complexes, are extremely efficient in the conjugate alkylation of α,β-unsaturated ketones. Since it is this reaction that is used most extensively in the synthesis of complex molecules, the development of mixed cuprates was an important synthetic advance.

Recently, organocopper chemistry has been developed far beyond these original boundaries. Thermally stable cuprates have been synthesized by using sterically hindered nontransferable ligands such as diphenylphosphide (Ph_2P) and dicyclohexylamide (Cy_2N) to prepare RCu(L)Li.[6] These reagents undergo typical organocuprate reactions, and are stable at 25° for over an hour. Even more impressive is the very hindered reagent RCu[P(tBu)$_2$]Li, which is stable in THF at reflux for several hours, but still enjoys organocuprate reactivity (Eq. 4.5).[7]

Treatment of copper(I) cyanide with two equivalents of an organolithium reagent produces the "higher order" cuprate $R_2Cu(CN)Li_2$, which is more stable than R_2CuLi, and is particularly effective for the reaction with normally unreactive secondary bromides or iodides, as well as for 1,4-addition to conjugated ketones and esters.[8] (The nature and even the existence of higher order cuprates is still a matter of controversy.[8c] Whatever the reagent is, its reactivity differs from that of simple diorganocuprates.) By taking advantage of the observation that mixed diaryl higher order cuprates do not scramble aryl groups at low (<–100°) temperatures, an efficient synthesis of unsymmetrical biaryls was developed (Eq. 4.6).[9]

Eq. 4.5

$[Cu[P(t\text{-}Bu)_2]_4 + 4RLi \longrightarrow RCu[P(t\text{-}Bu)_2]Li$

(x-ray)

good yields

R^1X

RR^1

Eq. 4.6

$ArLi + CuCN \xrightarrow{-78°} ArCu(CN)Li \xrightarrow[-125°C]{Ar'Li} [ArCuAr']CNLi_2 \xrightarrow[-125°C]{O_2} ArAr'$

78-90%

(>96% cross coupling)

Ar = Ph, oMeOPh, mMeOPh, pMeOPh, (2Cl, 4CF$_3$)Ph, 1 Naphth.
Ar' = oMeOPh, pMeOPh, oMePh, (3F,4Me)Ph

Addition of Lewis acids such as BF$_3$ to organocopper compounds increases their reactivity, and broadens their synthetic utility. The reagent RCu·BF$_3$ monoalkylates ketals with displacement of an alkoxy group, a very unconventional transformation (Eq. 4.7).[10] Optically active allylic acetals undergo S$_N$2' cleavage with very high diastereoselectivity (Eq. 4.8).[11]

Eq. 4.7

$R'Cu\cdot BF_3 + $

Eq. 4.8

79%
95% de

These Lewis acid modified cuprate reagents also efficiently add 1,4- to conjugated enones, and effective asymmetric induction in this process has been achieved in at least two different ways. The most widely studied system involves the use of unsaturated optically active sultam amides as substrates

(Eq. 4.9).[12] In most cases, both high yields and high enantiomeric excesses are observed. Similar efficiencies are seen with naphthyl substituted camphor esters (Eq. 4.10).[13]

Eq. 4.9

R'MgX + R'CuBF₃

Eq. 4.10

BuCuI•SiMe₃

93%   98% de

A potentially more general but less studied approach involves the use of optically active ligands on the metal, obviating the need for chiral auxiliaries on either the substrate or the reagent. A particularly successful example is shown in Eq. 4.11[14] although the generality of the approach has not yet been established.

Eq. 4.11

CuI, MeLi

L*, PhCH₃/THF

L* =

R(−) Muscone
89%
100% opt. purity

All of the organocopper chemistry discussed above relies upon the generation of the reactive organocopper species from either a Grignard reagent or an organolithium compound. This puts a serious limitation on the nature of the functional groups contained in the organocopper species, since they must be stable to organolithium reagents. Several solutions to the problem have recently been developed.

Reduction of CuCN·2LiBr with lithium naphthalenide at –100°C produces a highly active copper which can react with *functionalized* organic halides (I > Br >> Cl) to generate *functionalized* organocopper species in solution. Addition of electrophiles directly to these solutions produces alkylation products in excellent yield (Eq. 4.12).[15,16] By this process, ester groups, nitriles and both aryl and alkyl chlorides are inert, and cuprates containing these functional groups can be generated and transferred.

Eq. 4.12

Another route to functionalized organocopper reagents involves transmetallation to copper from less reactive organometallic reagents, particularly organozinc compounds. Organozinc halides can be made directly from organic halides and zinc metal, and many functional groups are quite stable to these organometallic reagents. Treatment of CuCN·2LiCl with functionalized organozinc halides produces functionalized organocuprates, which then undergo normal cuprate coupling processes, introducing *functionalized* R groups into the substrate. The current (1992) scope of this evolving process is shown in Figure 4.2.

Functionalized organic iodides undergo reaction with diethyl zinc followed by CuCN·2LiCl to give functionalized organocuprates that react in a similar fashion.[24] If the temperature is kept low (~ –100°C), many functional groups are stable even to organolithium reagents. Thus treatment of functionalized organic halides (FG = CN, CO$_2$tBu, NO$_2$, Cl, N$_3$) with butyllithium at

Figure 4.2 Reactions of Functionalized Organocopper Reagents

–100°C, followed by zinc chloride, followed by CuCN·2LiCl generates the functionalized organocopper species.[25]

Other functionalized organocopper complexes are also useful. α-Alkoxytin reagents are easily prepared by the reaction of trialkyl stannyllithium reagents with aldehydes. Although tin will not transmetallate to copper, the α-alkoxyalkyl group is readily cleaved from tin by butyllithium producing the α-alkoxyalkyllithium reagent. Addition of copper(I) cyanide followed by the typical array of electrophiles results in introduction of an α-alkoxyalkyl (Eq. 4.13).[26]

Eq. 4.13

$$Bu_3SnSnBu_3 + Li \longrightarrow 2Bu_3SnLi \xrightarrow[\text{(2) R'(+)}]{\text{(1) RCHO}} R\overset{OR'}{\underset{}{\underset{}{\diagup}}SnBu_3 \xrightarrow[\text{(3) E}^+]{\substack{\text{(1) BuLi} \\ \text{(2) CuCN}}} R\overset{OR'}{\underset{}{\diagup}E$$

Although dialkyl or alkyl heterocuprates are by far the most extensively used organocopper reagents, the simple organocopper complexes RCu or RCuMX perform much of the same chemistry, and are clearly superior for certain types of reactions, particularly the addition of $RCuMgX_2$ to terminal alkynes. R_2CuLi complexes often abstract the acidic acetylenic proton from terminal alkynes, and pure RCu or RCuLiX do not add at all. However, $RCuMgX_2$ complexes generated from Grignard reagents and copper(I) iodide cleanly add *syn* to simple terminal alkynes to produce vinylcuprates both regio- and stereospecifically ("carbocupration"). Note that the acidic acetylenic C–H bond is left intact. These vinyl complexes enjoy the reactivity common to organocopper species and react with a variety of organic substrates (Figure 4.3).[27] Functionalized cuprates generated as in Figure 4.2 are also "carbocuprate" alkynes, producing functionalized vinyl cuprates for use in synthesis (Eq. 4.14).[28] When carried out intramolecularly exocyclic alkenes are produced (Eq. 4.15).

Eq. 4.14

Eq. 4.15

55-76%

Figure 4.3 Carbocupration of Alkynes

4.3 σ-Carbon-Metal Complexes from Insertion of Alkenes and Alkynes into Metal-Hydrogen Bonds

The insertion of an alkene or alkyne into a metal hydrogen bond to form a σ-carbon-metal complex is a key step in a number of important processes, including catalytic hydrogenation and hydroformylation. This insertion is an equilibrium, with β-hydride elimination being the reverse step. With electron rich metals in low oxidation states [Rh(I), Pd(II)] the equilibrium lies far to the left, presumably because π-acceptor (electron withdrawing) olefin ligands stabilize these electron-rich metals more than the corresponding σ-donor alkyl ligands resulting from insertion (Eq. 4.16).

Eq. 4.16

The situation is the reverse with electron poor metals in high oxidation states (e.g., d^0 Zr(IV)). In these cases σ-donor alkyl ligands are stabilizing (and π accepting olefin ligands destabilizing) and the equilibrium lies to the right. From a synthetic point of view, the position of the equilibrium is not important, as long as the σ-carbon-metal complex formed by insertion is more reactive than the olefin-hydride complex. σ-Carbon metal complexes of both electron rich and electron poor metals, formed from metal hydrides and olefins, have been used in the synthesis of complex molecules.

An example of the former is the palladium(II) catalyzed cycloisomerization of enynes (Eq. 4.17).[29] Although the mechanism of this reaction is not yet known, it is likely to involve *in situ* generation of a palladium(II) hydride species, which "inserts" (hydrometallates) the alkyne to give a σ-vinyl palladium(II) complex with an adjacent alkene group. Coordination of this alkene, followed by insertion into the σ-vinyl palladium complex, forms a carbon carbon bond (the ring) and a σ-*alkyl* palladium(II) complex having two sets of β-hydrogens. β-Hydrogen elimination can occur in either direction, generating the cyclic diene and regenerating the palladium(II) hydride species to reenter the catalytic cycle. The direction of β-elimination can be influenced by the addition of ligands (Eq. 4.18), although the reason for this is not entirely clear.

Eq. 4.17

The reaction tolerates a range of functional groups (Eq. 4.19), as well as structural complexity in the substrate (Eq. 4.20), (Eq. 4.21) and (Eq. 4.22). When carried out in the presence of an appropriate reducing agent, the ultimate σ-alkylpalladium(II) complex can be reduced prior to β-elimination, giving monoenes rather than dienes (Eq. 4.23).

Eq. 4.18

Pd(OAc)$_2$ (80%)   1:16
Pd(OAc)$_2$/2 PPh$_3$ (73%)   1:2.9

Eq. 4.19

R = COCH$_3$, TMS, CH$_2$OCH$_3$, CH$_2$CH$_2$OCH$_3$, CH$_3$, OEt, CO$_2$Me

Eq. 4.20

Eq. 4.21

Eq. 4.22

Eq. 4.23

Multiple insertions are possible, provided β-hydrogen elimination is slower than the next insertion, a feature achieved by careful selection of substrate. The example in Eq. 4.24 is illustrative. Insertion is a *cis* process, and β-elimination requires that the metal and hydrogen achieve a *syn* coplanar relationship. In Eq. 4.24, a *cis* insertion leads to σ-alkyl palladium(II) complex which has no β-hydrogen *syn* coplanar, making β-elimination disfavored. Instead, a suitably situated alkene undergoes insertion, and the resulting σ-alkyl palladium(II) complex can then undergo β-elimination. The ultimate example of this is shown in Eq. 4.25 in which all but the final center formed are quaternary, and have no β-hydrogens.[30] This process has been used to generate 1,3,5-hexatrienes which undergo a facile electrocyclic reaction to produce tricyclic material (Eq. 4.26).[31] Note that in none of these cases could σ-carbon-metal complexes be detected, since they were present only in low concentrations because the insertion equilibrium lay far to the left.

Eq. 4.24

Eq. 4.25

Eq. 4.26

The completely opposite situation obtains for the electron poor, d^o, Zr(IV) complex $Cp_2Zr(H)Cl$, which reacts under mild conditions with a variety of alkenes to give stable, isolable σ alkyl complexes of the type $Cp_2Zr(R)Cl$ (Eq. 4.27).[32]

Eq. 4.27

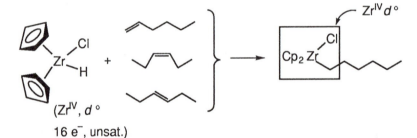

This reaction is remarkable in several respects. The addition occurs under mild conditions, and the resulting alkyl complexes are quite stable. Regardless of the initial position of the double bond in the substrate, the Zr ends up at the sterically least-hindered accessible position of the olefin chain. This rearrangement of the zirconium to the terminal position occurs by a Zr–H elimination, followed by a readdition that places Zr at the less-hindered position of the alkyl chain in each instance (Eq. 4.28). This migration proceeds because of the steric congestion about Zr caused by the two cyclopentadienyl rings. The order of reactivity of various olefins toward hydrozirconation is terminal olefins > *cis* internal olefins > *trans* internal olefins > exocyclic olefins

> cyclic olefins and terminal > disubstituted > trisubstituted olefins. Tetrasubstituted olefins and trisubstituted cyclic olefins fail to react. Finally, 1,3 dienes add ZrH to the less-hindered double bond to give γ,δ-unsaturated alkylzirconium complexes.

Eq. 4.28

Alkynes also react with $Cp_2Zr(H)Cl$, adding Zr–H in a *cis* manner. With unsymmetric alkynes, addition of Zr–H gives mixtures of vinyl Zr complexes with the less-hindered complex predominating. Equilibration of this mixture with a slight excess of $Cp_2Zr(H)Cl$ leads to mixtures greatly enriched in less-hindered complex. The good news is that Cp_2ZrHCl is a metal hydride which readily adds to a wide variety of alkenes and alkynes to give a metal-alkyl species, in principle making alkenes and alkynes sources of alkyl groups in synthesis. The bad news is that Cp_2ZrRCl is Zr(IV), d^0 and undergoes very little useful organic chemistry.

Zirconium(IV) alkyls do undergo two useful reactions. They are readily cleaved by electrophiles, with retention of configuration, to give straight-chain alkyl halides (Eq. 4.29) or alcohols (Eq. 4.30)[33] even from mixtures of alkene isomers. (Zirconium(IV) alkenyls undergo electrophilic cleavage with retention of alkene geometry.) In addition, they insert carbon monoxide, giving stable acylzirconium(IV) complex which can be cleaved by acid to give aldehydes (Eq. 4.31).[34] However it was not until the development of efficient transmetallation processes that hydrozirconation achieved real utility in organic synthesis.

Eq. 4.29

Eq. 4.30

Eq. 4.31

4.4 σ-Carbon-Metal Complexes from Transmetallation/Insertion Processes

Transmetallation (transfer of an R group) from main group metals to transition metals is extensively exploited in organic synthesis, notwithstanding the fact that the process is poorly understood (Chapter 2). Its utility lies in the ability to use main group organometallic chemistry to "activate" organic substrates (e.g., hydroboration of alkenes, direct mercuration of aromatics), then transfer that substrate to transition metals to take advantage of their unique reactivity (e.g., oxidative addition, insertion). One such useful class of reaction is transmetallation/insertion (Eq. 4.32).

Eq. 4.32

As mentioned above, although alkenes and alkynes are readily hydrozirconated, the resulting σ-alkylzirconium(IV) species have only minimal reactivity. However, they will undergo transmetallation to nickel(II) complexes, and the resulting organonickel complexes undergo 1,4-addition to conjugated enones (Eq. 4.33).[35] Multiple transmetallations increase the utility of the process. For example, in Eq. 4.34,[36] hydrozirconation is used to activate the alkyne, but the σ-vinylzirconium(IV) species cannot directly transmetallate to copper. However, organolithium reagents will transmetallate to give

Eq. 4.33

Eq. 4.34

vinyllithium which in turn transmetallates to copper, which then does 1,4-additions to an enone (Eq. 4.35).[37] Alternatively, Cp$_2$ZrCl$_2$ catalyzes the "carboalumination" of alkynes[38] to give vinylalanes which will transmetallate directly to copper (Eq. 4.36).[39] In this way each metal is used to confer its unique reactivity on the system. The combination can be powerful.

Eq. 4.35

Eq. 4.36

Transmetallation from mercury, thallium and tin to palladium(II) is also a powerful synthetic technique. Aromatic compounds undergo direct, electrophilic mercuration or thallation and, although these main group organometallics have limited reactivity, they will transfer their aryl groups to palladium(II), which has a very rich organic chemistry. When transmetallation/insertion chemistry is used, the reaction is, unfortunately, stoichiometric in palladium, since Pd(II) is required for insertion, and Pd(0) is produced in the last step (efficient reoxidation has not been achieved) (Eqs. 4.37, 4.38[40] and 4.39[41]). However, this can still be a useful process, since Pd(0) can be recovered and reoxidized in a separate step. In the special case of an allylic chloride as the alkene (Eq. 4.40),[42] the reaction *is* catalytic in Pd(II) since, in this case, β-Cl elimination to give PdCl$_2$, rather than β-hydride elimination to give Pd(0) and HCl, is more facile. Transmetallation from arylthallium(III) complexes is inefficiently catalytic (Eq. 4.41).[43] In this case, Tl(III) oxidizes Pd(0) to Pd(II). Transmetallation from tin has been used to synthesize arylglycosides (Eq. 4.42).[44] Even arylmanganese(I) complexes transmetallate to Pd(II), permitting olefination (Eq. 4.43).[45]

Eq. 4.37

Eq. 4.38

R = H, CO$_2$Me, CO$_2t$Bu,

Eq. 4.39

50-70%

Eq. 4.40

Eq. 4.41

$$+ \quad Pd(0) \xrightarrow{Tl^{III}} Pd(II) + Tl(I)$$

30%

Eq. 4.42

60%

Eq. 4.43

Z = CN, CO$_2$Me, CHO

4.5 σ-Carbon-Metal Complexes from Oxidative Addition/Transmetallation Sequences

One of the most general approaches to the formation of σ-carbon metal complexes is the oxidative addition (Chapter 2) of organic halides to low valent transition metals, most commonly palladium(0) and nickel(0) ((a) in Figure 4.4). The reaction is usually restricted to aryl and vinyl halides, or at least halides lacking β-hydrogens, since β-hydride elimination is very fast, limiting what can be done with the σ-carbon metal complex. With vinyl halides, retention of stereochemistry is observed. The order of reactivity is I > OTf > Br >>> Cl, such that chlorides are rarely reactive enough to participate. From the viewpoint of the substrate, the metal is being oxidized and the substrate reduced. Thus electron deficient substrates are more reactive than electron rich substrates. The resulting (unstable) σ-carbon metal halide complex can undergo a variety of useful reactions. Treatment with a main group organometallic results in transmetallation, (b) producing a diorganometal complex, which undergoes rapid reductive elimination (c) to give coupling (carbon-carbon bond formation) with regeneration of the low valent metal catalyst (Figure 4.4). Because reductive elimination is faster than β-hydride elimination, the main group organometallic *may* contain β-hydrogens, and is not restricted to sp^2 carbon groups. Transmetallation and reductive elimination usually occur with retention of stereochemistry.

Because CO insertion into metal carbon σ-bonds is facile at temperatures as low as –20°C, carrying out the oxidative addition under an atmosphere of CO results in the capture of the σ-carbon metal complex by CO to produce a σ-acyl species (d). Transmetallation (e) to this complex results in carbonylative coupling (f) (Figure 4.4). Both of these processes have been extensively used in synthesis.

One of the earliest examples of oxidative addition/transmetallation is the nickel- or palladium-catalyzed Grignard reaction. (Although Ni(II) or Pd(II) complexes are the most common catalyst precursors, the active catalyst species is the metal(0) formed by reduction of the metal(II) complex by the Grignard reagent.) Nickel phosphine complexes catalyze the selective cross-coupling of Grignard reagents with aryl and olefinic halides (Eq. 4.44).[46] The

$$M = Mg, Zr, \underline{Zn}, Sn, B, Al, Si, Cu$$

Figure 4.4 Oxidative Addition/Transmetallation

reaction works well with primary and secondary alkyl, aryl, alkenyl, and allylic Grignard reagents. A wide variety of simple, fused, or substituted aryl and alkenyl halides are reactive substrates, but simple alkyl halides fail to react. With simple aryl and alkyl Grignard reagents, the chelating diphosphine $Ph_2P(CH_2)_2PPh_2$ complex of nickel chloride is the most effective catalyst, whereas the $Me_2P(CH_2)_2PMe_2$ complex is most suitable for alkenyl and allylic Grignard reactions, and the *bis* Ph_3P complex is best for reactions of sterically hindered Grignards or substrates.

Eq. 4.44

$$R = alkyl, aryl, hetaryl, vinyl$$

This process is particularly useful for the coupling of heterocyclic substrates (Eq. 4.45).[47] In this case the Grignard reagent, the halide, or both can be heterocycles.

Eq. 4.45

Configurationally unstable secondary alkyl Grignard reagents can be coupled to a variety of aryl and vinyl halides with high asymmetric induction using (aminoalkylferrocenyl)phosphines having both carbon centered and planar chirality (Eq. 4.46).[48] This is a *kinetic resolution* of the racemic (but rapidly equilibrating) Grignard reagent — that is, one enantiomer of the Grignard reagent reacts with the optically active phosphine metal(II) σ-carbon complex faster than does the other enantiomer, and the Grignard reagent equilibrates (inverts) faster than it couples.

Eq. 4.46

By far the most extensively developed oxidative addition/transmetallation chemistry involves palladium(0) catalysis, and transmetallation from Zn, Zr, B, Al, Sn, Si and Cu. There is an overwhelming amount of literature on the subject, but all examples have common features. A very wide range of palladium catalysts or catalyst precursors have been used (Figure 4.5) and, in many cases, catalyst choice is not critical. These catalysts include preformed, stable palladium(0) complexes such as Pd(PPh₃)₄ ("tetrakis") or Pd(dba)₂/Pd₂(dba)₃·CH₃Cl plus phosphine; *in situ* generated palladium(0) phosphine complexes, made by reducing palladium(II) complexes in the presence of phosphines; and "naked" or ligandless palladium(0) species, including, in some cases, palladium on carbon. (It should be noted that palladium(II) is very readily reduced to palladium(0) by alcohols, amines, CO, olefins, phosphines, and main group organometallics.) Phosphine ligands are normally required for oxidative addition of bromides, but they suppress the reaction with aryl iodides, thus discrimination between halides is efficient. In these coupling processes, almost invariably it is the transmetallation step that is rate limiting, and is the least understood. It is important that *both* metals ultimately benefit energetically from the transmetallation step and, depending on the metal, additives are sometimes required.

Figure 4.5   Preparation of Palladium(0) Complexes

Organozinc halides are among the most efficient main group organometallics in transmetallation reactions to palladium (Eq. 4.47),[49] while organolithium reagents are among the least. However, treatment of an organolithium reagent with one equivalent of anhydrous zinc chloride in THF generates, *in situ*, an organozinc halide which couples effectively (Eq. 4.48[50] and Eq. 4.49[51]). This allows the use of ligand-directed metallation to provide coupling partners in this process. Another attractive feature of organozinc halides is that a wide variety of *functionalized* reagents are available directly from functionalized halides (see Eqs. 4.12 to 4.15) permitting the introduction of carbonyl groups, nitriles, halides, and other species into these coupling reactions.

Eq. 4.47

Eq. 4.48

Eq. 4.49

σ-Vinylzirconium(IV) complexes from hydrozirconation of *terminal* alkynes also transmetallate efficiently to palladium(II)-carbon complexes and are effective coupling partners in Pd(0) catalyzed oxidative addition/transmetallation processes (Eq. 4.50[52] and Eq. 4.51[53]). However, the corresponding σ-vinylzirconium(IV) complexes from hydrozirconation of *internal* alkynes fail entirely! Addition of zinc chloride to these reactions promotes this process, probably via a Zr → Zn → Pd transmetallation sequence (Eq. 4.52).[54] This, again, illustrates the efficacy of zinc organometallics in these coupling processes.

Eq. 4.50

Eq. 4.51

Eq. 4.52

Hydroboration is potentially a very attractive way to bring alkynes and alkenes into oxidative addition/transmetallation cycles, since the process has been the subject of long term, intense development and a truly staggering number of cases have been studied. However B → Pd transmetallation proved difficult until it was realized that the process could be facilitated by the addition of base, probably via the borate complex (Eq. 4.53).[55] This reaction has very wide applicability and is responsible for one of the most impressive couplings of this class yet reported (Eq. 4.54).[56] The reaction can also be carried out intramolecularly in a combined hydroboration/oxidative addition/transmetallation sequence (Eq. 4.55).[57]

Eq. 4.53

Eq. 4.54

75%

Eq. 4.55

Transmetallation has become increasingly important in the synthesis of highly functionalized aryl compounds with the combination of functional group directed arene metallation ("ortho lithiation")/boronation and oxidative addition/transmetallation.[58] Functional group directed metallation of arenes has been very extensively developed. Quenching these lithiated arenes with trimethyl borate produces organoboron compounds which undergo efficient transmetallation to palladium (Eq. 4.56,[59] Eq. 4.57,[60] Eq. 4.58,[61] and Eq. 4.59[62]).

Eq. 4.56

Eq. 4.57

Eq. 4.58

Eq. 4.59

Transmetallation from tin to palladium is one of the most extensively utilized and highly developed palladium catalyzed processes in organic synthesis,[63] primarily because of the ease of preparation and handling of a very wide variety of highly functionalized tin reagents, along with the facility with which they transmetallate to palladium. The inherent toxicity of tin compounds and the difficulty in removing tin by-products from reactions have done little to dampen this enthusiasm.

The rate, and ease of transfer from tin to palladium is:

$$RC\equiv C > RCH=CH > Ar > \text{\includegraphics{allyl}} , ArCH_2 > > C_nH_{2n+1}, RCHOR$$

Vinyl groups transfer with retention of olefin geometry, and allyl groups with allylic transposition. Because simple alkyl groups transfer very slowly, trimethyl or tributylstannyl compounds can be used to transfer a single R group. Transmetallation is almost invariably the rate limiting step in these processes, and relatively unreactive tin reagents require high (>100°C) reaction temperatures and long reaction times.

In principle, any substrate that will undergo oxidative addition with palladium(0) can be coupled to any organotin group that will transfer. Among the earliest studied substrates were acid chlorides, which are readily converted to ketones (Eq. 4.60).[64] Note that a methyl group can be transferred if it is the only group on tin and conditions are vigorous enough (Eq. 4.61).[65] Macrocycles can be made using an intramolecular version of this process (Eq. 4.62).[66] (A carbon monoxide atmosphere is required to prevent competitive decarbonylation (Chapter 5) of the acid chloride.) Acid chlorides are reduced to aldehydes when tributyltin hydride is used (Eq. 4.63).[67]

Eq. 4.60

Eq. 4.61

Eq. 4.62

n = 13, 9, 7, 5, 4

R = H, Me, nC_6

Eq. 4.63

Vinyl halides couple efficiently with acetylenic (Eq. 4.64)[68] and simple vinyl (Eq. 4.65) tin compounds, with retention of olefin geometry for both partners.[69] The reaction has been used to append side chains to functionalized macrocycles (Eq. 4.66),[70] and intramolecularly to form macrocyclic lactones (Eq. 4.67).[71]

Eq. 4.64

Eq. 4.65

Eq. 4.66

Eq. 4.67

Zearalenone 54%

The synthetic utility of this coupling process was dramatically enhanced when conditions were found to make aryl and vinyl triflates[72] and fluorosulfonates[73] participate, since this allowed phenols and ketones (via their enol triflates or fluorosulfonates) to serve as coupling substrates. As with other substrates, extensive functionality is tolerated in both partners (Eq. 4.68[74], Eq. 4.69[75], and Eq. 4.70[76]). In many, but not all, cases reactions with triflates require the addition of LiCl or $ZnCl_2$ to go smoothly. It has been postulated (but not demonstrated) that the formation of the more stable R_3SnCl rather than the less stable R_3SnOTf is responsible for this observation. Even electron rich, sterically hindered aryl triflates will couple, provided >10% catalyst is used, and triphenyl phosphine is added (Eq. 4.71).[77]

Eq. 4.68

Eq. 4.69

Eq. 4.70

Eq. 4.71

$$R^3 = Me,\ Ph,\ PhC\equiv C\!-\!,\ Ar$$

Carbon monoxide insertion into metal-carbon σ-bonds is a very facile and rapid process occurring at temperatures as low as –20°C, while transmetallation is usually quite slow. This permits the trapping of the initial oxidative adduct σ-carbon metal complex by CO, producing a σ-acyl complex. Transmetallation to this species, followed by reductive elimination, leads to carbonylative coupling (Figure 4.4, Eq. 4.72).[78] Vinyl triflates also undergo carbonylative coupling efficiently (Eq. 4.73)[79] and the reaction has been carried out intramolecularly to form macrocyclic lactones (Eq. 4.74).[80]

Eq. 4.72

Transmetallation from silicon to palladium is not a facile process, and vinyl silanes can often be carried along as unreactive functional groups in other transmetallation/coupling schemes (e.g., Eq. 4.73). However in the presence of a fluoride source, such as tetrabutylammonium fluoride (TBAF) or (tris(diethylamino)sulfonium difluorotrimethylsilicate) (TASF), to aid in Si–C bond cleavage and to provide stable silicon products, oxidative addition/transmetallation from silicon proceeds smoothly with a wide variety of substrates (Eq. 4.75).[81] In contrast to other main group organometallics, both *cis* and *trans* vinyl silanes couple with aryl halides to give *trans* products.[82]

Eq. 4.73

87%

Eq. 4.74

n = 4-8
40-70%

Eq. 4.75

RX + R'SiMe$_3$ → RR'

2% ⟨⟨ PdCl$_2$ cat

HMPA or THF/P(OEt)$_3$
50° 1 eq TASF   60-90%

RX = ArI, ... , ArOTf, ...OTf

R' = ...

Preformed organocuprates, R_2CuLi, transmetallate to palladium only slowly, and most decompose at temperatures well below that required for transmetallation. However, terminal alkynes couple to aryl and vinyl halides upon treatment with palladium catalyst, an amine and copper(I) iodide, in a process that almost certainly involves oxidative addition/transmetallation from copper. Although the mechanism has not been studied, a likely one is shown in Eq. 4.76. This reaction has been known for a long time, but only became appreciated with the advent of the ene-diyne antibiotic, antitumor agents such as calicheamycin or esperamycin.[83] It turns out that this is an excellent reaction, which retains the stereochemistry of the vinyl halide (Eq. 4.77[84] and Eq. 4.78[85]). The amine is apparently required, both to act as a base for copper acetylide production and to reduce palladium(II) precatalysts to palladium(0) when these are used (Eq. 4.79[86] and Eq. 4.80[87]). The reaction tolerates a wide range of functionality (Eq. 4.81[88] and Eq. 4.82[89]).

Eq. 4.76

Eq. 4.77

Eq. 4.78

Eq. 4.79

R = *n*Pr, *n*Bu, Ph
R′ = F, Ph, CF₃, (*i*PrO)₂P(O)

Eq. 4.80

Eq. 4.81

Eq. 4.82

4.6 σ-Carbon-Metal Complexes from Oxidative Addition/Insertion Processes (the Heck Reaction)

Metal carbon σ-bonds are readily formed by oxidative addition processes, and carbon monoxide and (less easily) alkenes readily insert into metal carbon σ-bonds. These two processes form the basis of a number of synthetically useful transformations. Because of the very wide range of organic substrates that undergo oxidative addition, carbon monoxide insertion offers one of the best ways to introduce carbonyl groups into organic substrates. The general process is shown in Figure 4.5. As usual, it is limited to substrates lacking β-hydrogens, since β-hydride elimination often occurs under the conditions required for oxidative addition. Thus, vinyl, aryl, benzyl, and allyl halides, triflates, and fluorosulfonates[90] are readily carbonylated. Pressure equipment is usually not required, since insertion occurs readily at one atmosphere pressure. A base is required to neutralize the acid produced from the nucleophilic cleavage of the σ-acyl complex. When palladium(II) catalyst precursors are used, rapid reduction to palladium(0) by carbon monoxide ensues. Some ligand, usually triphenyl phosphine, is required both to facilitate oxidative addition with bromides and to prevent precipitation of metallic palladium. Vinyl triflates have been extensively used (Eq. 4.83[91] and Eq. 4.84[92]). Because they are directly available from ketones, this provides a convenient method for the conversion of this common functional group to conjugated esters.

Figure 4.5 Oxidative Addition/Insertion

Eq. 4.83

Eq. 4.84

Intramolecular versions are efficient and provide lactones (Eq. 4.85),[93] lactams (Eq. 4.86)[94] and even β-lactams (Eq. 4.87).[95] These carbonylations can also be carried out on a large scale without loss of efficiency (Eq. 4.88).[96] When amino alcohols are used to trap palladium acyl complexes, oxazolines (Eq. 4.89)[97] and benzoxazoles (Eq. 4.90)[98] can be prepared from a variety of aryl and vinyl halides and triflates. Even the oxygen of an appropriately situated enolate can trap the σ-acyl palladium complex to produce isocoumarins (Eq. 4.91).[99] These examples are only a very narrow sampling of a very general reaction with broad applications.

Eq. 4.85

Eq. 4.86

Eq. 4.87

Eq. 4.88

(89% on 180g scale)

Eq. 4.89

75-95%

Eq. 4.90

60-80%

Eq. 4.91

80-90%

Oxidative addition/olefin insertion sequences are even more useful for synthesis, because a greater degree of elaboration is introduced. The general process, known as the Heck reaction,[100] is shown in Figure 4.6. In this case, it is normally the insertion step which is most difficult and it is the structural features on the alkene which impose the most limitations. *All* substrates subject to oxidative addition can, in principle, participate in this process, although, again, β-hydrogens are not tolerated because β-elimination competes with insertion.

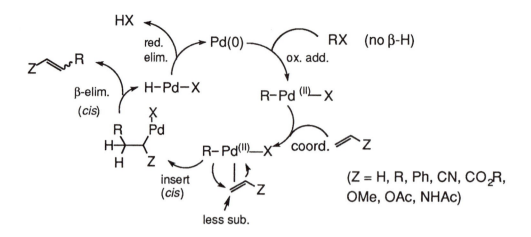

Catalysts: L_4Pd, $Pd(dba)_2$, $PdCl_2L_2$ + DIBAL, $Pd(OAc)_2$ + red. agent
(CO, CH_2=CH_2, R_3N, R_3P, ROH)

Figure 4.6 Oxidative Addition/Olefin Insertion

Typically, the Heck reaction is carried out by heating the substrate halide or triflate, the alkene, a catalytic amount of palladium(II) acetate, and an excess of a tertiary amine in acetonitrile. The tertiary amine rapidly reduces Pd(II) to Pd(0) by coordination/β-elimination/reductive elimination (Eq. 4.92).[101] With aryl iodides as substrates, the addition of phosphine is not required, and in fact, inhibits the reaction. Even palladium on carbon will catalyze Heck reactions of aryl iodides.[102] In contrast, with bromides as substrates, a hindered tertiary phosphine like tri-o-tolylphosphine is required to suppress quaternization. Recently, considerably milder, more efficient conditions for the Heck reaction have been developed. These involve using polar solvents such as DMF with added quaternary ammonium chlorides to promote the reaction.[103] The soluble quaternary ammonium chloride keeps the concentration of soluble chloride high, and halide ions stabilize and activate palladium(0) complexes during these reactions.[104] For industrial applications, related conditions which result in very high catalyst turnover have been developed.[105]

Eq. 4.92

$$R_2NCH_2R' \ + \ PdX_2 \ \rightleftharpoons \ \left[\begin{array}{c} H \\ | \\ R_2N-C-R \\ | \quad | \\ Pd \ H \\ | \\ X \end{array} \right]^+ \ \xrightarrow[\ X^-\ ]{\beta\text{-el.}} \ \left[R_2\overset{+}{N}=CHR \right] X^- \ + \ \overset{H}{PdX} \ \longrightarrow \ Pd(0) \ + \ HX$$

A very wide range of alkenes will undergo this reaction, including electron deficient alkenes, like acrylates, conjugated enones and nitriles, unactivated alkenes like ethylene and styrene, and electron rich alkenes like enol ethers and enamides. With electron poor and unactivated alkenes the insertion step is sterically controlled and insertion occurs to place the R group at the less substituted position. The R group is delivered from the same face as the metal (*cis* insertion). Substituents on the β-position of the alkene inhibit insertion and, in intermolecular reactions, β-disubstituted alkenes are unreactive. The β-hydrogen elimination is also a *cis* process (Pd and H must be *syn* coplanar).

With electron rich alkenes, the regioselectivity of insertion is more complex.[106,107] With cyclic enol ethers and enamides, arylation α- to the heteroatom is always favored, but with acyclic systems mixtures are often obtained. Bidentate phosphine ligands favor α-arylation, as do electron rich aryl halides, while electron poor aryl halides tend to result in β-arylation. The regioselectivity of *intra*molecular processes is also difficult to predict, and seems to depend more on ring size than alkene substitution (see below).

There are literally thousands of examples of the use of the Heck reaction in organic synthesis, and many complex systems have been studied. Functional group compatibility is excellent and isolated yields are often very high (Eq. 4.93,[108] Eq. 4.94,[109] and Eq. 4.95[110]). Intramolecular versions are even more useful (Eq. 4.96[111] and Eq. 4.97)[112] although the regioselectivity is sometimes surprising (Eq. 4.98 and Eq. 4.99[113]). By using chiral ligands, asymmetric induction can be efficient (Eq. 4.100[114] and Eq. 4.101[115]).

Eq. 4.93

X = CO₂Me, COMe, Ph, CN

Eq. 4.94

Eq. 4.95

The olefin insertion step in the Heck reaction generates a new metal-carbon σ-bond which, in principle, could undergo any of the reactions of metal-carbon σ-bonds (Figure 4.1), if only β-hydrogen elimination did not occur so readily. If β-elimination could be suppressed or if a process even more favored than β-elimination (such as CO insertion) could be arranged, then the metal could be used more than once, and several new bonds could be formed, in sequence, in a single catalytic process. This is, indeed, possible, and is the focus of a great deal of current (1992) effort.

Eq. 4.96

89%

Eq. 4.97

91%

Eq. 4.98

89%

Eq. 4.99

74%

Eq. 4.100

40-60%

>96% ee

Ar = Ph, pClPh, mClPh, o ClPh,
pAcPh, pNCPh,
p MeOPh, 2-Naphth.

Eq. 4.101

$$\text{PdCl}_2 / (R)\text{-BINAP}$$
$$\text{Ag}_2\text{CO}_3$$

40-67%
76% ee

For example, the reaction in Eq. 4.102[116] involves an oxidative addition CO insertion/olefin insertion/CO insertion/cleavage process that forms three carbon-carbon bonds per turn of catalyst. This process works because the initial oxidative adduct has no accessible β-hydrogens, and CO insertion is faster than olefin insertion for σ-alkyl palladium complexes but slower for σ-acyl complexes. The major challenge in the system was to make the final CO insertion into a σ-alkyl palladium(II) complex, which does have β hydrogens, competitive with β-elimination. This was achieved by using a relatively high pressure of CO. Indeed, under one atmosphere of CO, β elimination was the main course of the reaction. This downhill sequence of events is critical in these multiple cascade reactions, and the requisite fine balance cannot always be achieved, as evidenced by the failure when an alkyne insertion is required to precede the much more facile CO insertion (Eq. 4.103).[117] However, this same reaction was quite successful when an intermolecular alkene insertion had to compete with an intramolecular alkyne insertion (Eq. 4.104). Since these two insertions are of comparable energy, and since the alkyne insertion is entropically favored, the desired sequences of steps occurred without complication.[117] In general, as long as there is a vinyl halide to start the process, and appropriately situated olefinic or acetylenic groups to carry the insertion cascade without suffering β-hydride elimination, truly remarkable transformations can occur (Eq. 4.105,[118] Eq. 4.106, Eq. 4.107,[119] and Eq. 4.108![120]). The process can even be truncated (Eq. 4.109)[121] or initiated (Eq. 4.110)[122] by intermolecular steps.

Eq. 4.102

Eq. 4.103

30 : 70
60%

Eq. 4.104

72%

Eq. 4.105

76%

Eq. 4.106

86%

Eq. 4.107

3% Pd(PPh₃)₄

Et₃N, MeCN
reflux, 12h

76%

E = CO₂Et

Eq. 4.108

Eq. 4.109

81%

Eq. 4.110

60%

In all of the above cases, the last step of the cascade was a β-hydrogen elimination. However, if β-elimination is not possible (for example, when the last insertion involves an alkyne) the resulting σ-vinyl palladium(II) complex can be trapped by transmetallation (Eq. 4.111)[123] or nucleophilic attack (Eq. 4.112).[124] Almost endless variations of these themes are possible, and most are likely to be attempted.

Eq. 4.111

Eq. 4.112

4.7 σ-Carbon-Metal Complexes from Ligand Directed Cyclometallation Processes

Palladium(II) salts are reasonably electrophilic, and under appropriate conditions (electron rich arenes, Pd(OAc)$_2$, boiling acetic acid) can directly palladate arenes via an electrophilic aromatic substitution process. However this process is neither general nor efficient, and thus has found little use in organic synthesis. Much more general is the ligand directed orthopalladation shown in Eq. 4.113.[125] This process is very general, the only requirement being a lone pair of electrons in a benzylic position to precoordinate to the metal, and conditions conducive to electrophilic aromatic substitution (usually HOAc solvent). The site of palladation is always ortho to the

directing ligand and, when the two ortho positions are inequivalent, palladation results at the sterically less congested position, regardless of the electronic bias. (This is in direct contrast to *o*-lithiation, which is electronically
controlled and will usually occur at the more hindered position on rings having several electron donating groups.) The resulting, chelate stabilized, σ-
arylpalladium(II) complexes are almost invariably quite stable, are easily isolated, purified and stored, and are quite unreactive. This has limited their use
in synthesis and has prevented successful catalytic processes involving *o*-
palladation.

Eq. 4.113

These orthopalladated species *can* be made to react like "normal" σ-
aryl metal complexes by decomplexing the ortho ligand, usually accom-
plished by treating the complex with the substrate and a large excess of triethyl amine, to compete with the ortho ligand. Under these conditions alkene
insertion (Eq. 4.114)[126] and CO insertion (Eq. 4.115)[127] are feasible. Although
catalysis has not yet been achieved, the potential power of ligand directed
metallation is illustrated in Eq. 4.116,[128] in which an unactivated methyl
group was palladated solely because it had the misfortune of being within the
range of a ligand directed palladation.

Eq. 4.114

Eq. 4.115

Eq. 4.116

4.8 σ-Carbon-Metal Complexes from Reductive Cyclodimerization of Alkenes and Alkynes

Low valent metals from the two extremes of the transition series (Ni, Ti, Zr) undergo reaction with alkenes and alkynes to form five-membered metallacycles in which the unsaturated substrates are formally reductively coupled, and the metal is formally oxidized (Eq. 4.117). These metallacycles can be converted to a number of interesting organic compounds.

Eq. 4.117

The most extensively studied complex to affect this reaction is "zirconocene," Cp_2Zr, an unstable, unsaturated, d^2 Zr(II) complex. It can be generated in many ways (Figure 4.7) but the most convenient is to treat Cp_2ZrCl_2 with butyllithium and allow the unstable dibutyl complex warm to above –20°C in the presence of substrate. This generates a zirconium-butene complex which can exchange with substrate alkene or alkyne and promote

Figure 4.7 Formation of "Zirconocene"

the cyclodimerization. The zirconium butene complex can be considered either as a Zr(II)-alkene complex or a Zr(IV) metallacyclopropane complex. These are the two extreme bonding forms for metal alkene complexes, with the metallacyclopropane form denoting complete two electron transfer from the metal to the olefin. For the zirconium butene complex, the metallacyclopropane form is probably the more accurate representation, based on its reaction chemistry and the isolation of a stable PMe₃ complex. Given that "Cp₂Zr" in fact reductively dimerizes alkenes and alkynes (Eq. 4.117) (2 e⁻ transfer from Zr(II) to alkenes), it is not surprising that the zirconium butene complex also involves extensive electron transfer to the alkene.

The five-membered zirconacycles formed by cyclodimerization of alkenes can be transformed into a variety of interesting products. Protolytic cleavage produces the alkane, iodine cleavage the diiodide, and treatment with carbon monoxide and iodine, the cyclopentanone (Eq. 4.118).[129] Cyclodimerization of alkynes produces zirconacyclopentadienes, which undergo electrophilic cleavage to produce a number of unusual heterocycles (Eq. 4.119).[130]

Eq. 4.118

Eq. 4.119

E = SO, S, PhP, Me$_2$Sn, Se, GeCl$_2$, PhAs, PhSb

Intramolecular enyne and diyne cyclizations are also effective, forming mono or bicyclic compounds, depending on the method of cleavage (Eq. 4.120).[131] A range of functional groups is tolerated, and the reaction is highly stereoselective (Eq. 4.121[132] and Eq. 4.122[133]). When enantiomerically pure starting materials were used, ≥98% ee in the product was observed. In this manner, precursors to dendrobine were synthesized very efficiently from simple starting materials (Eq. 4.123).[134]

Eq. 4.120

good yields
many examples

Eq. 4.121

Eq. 4.122

Treatment of Cp$_2$Zr(Me)Cl with amide anions produces unstable zirconium-imine complexes, probably by a sequence of Cl$^-$ displacement, β-hydride elimination and reductive elimination (Eq. 4.124). These insert alkynes to give azazirconacyclopentenes which are cleaved by acid to give allyl amines (Eq. 4.124),[135] or by high pressure carbon monoxide to give pyrroles (Eq. 4.125).[136] An intramolecular version of this shows that these same complexes are probably directly available from imines and "zirconocene" (Eq. 4.126).[137]

Eq. 4.123

Eq. 4.124

60-80%

Eq. 4.125

40-70%

Eq. 4.126

70-80%

In a closely related process, zirconocene dichloride catalyzes the reaction between alkenes and Grignard reagents to produce new Grignard reagents. The intramolecular version is the easiest to follow (Eq. 4.127).[138] Treatment of α,ω-dienes with butylmagnesium chloride in the presence of catalytic amounts of Cp_2ZrCl_2 produces cyclized, *bis* Grignard reagents, which can be further functionalized. The reaction is thought to proceed as in Figure 4.7. Treatment of zirconocene dichloride with butyl Grignard reagent produces the reactive zirconacyclopropane/zirconocene butene complex (a). This reacts with the diene to produce the zironacyclopentane (b). Once this occurs, the zirconacyclopentane is cleaved by the Grignard reagent to generate the *bis* Grignard and regenerate the $Cp_2Zr(Bu)_2$. With monoolefins, Cp_2ZrCl_2 catalyzes the addition of EtMgCl to alkenes to give homologated Grignard reagents (Eq. 4.128).[139] The process is similar to that with diolefins, except in this case, the Grignard provides one "alkene", and the cleavage gives a mono rather than a *bis* Grignard reagent.

Eq. 4.127

Eq. 4.128

Perhaps most synthetically interesting is the reaction with allylic alcohols and ethers (Eq. 4.129).[140] Here it is clear that the allylic oxygen has both stereochemical and regiochemical consequences, indicating complexation to zirconium along the course of the reaction. With allylic alcohols, the zirconium reagent is delivered to the *same* face as the OH group, while with allylic ethers it comes from the opposite face, implying that OH complexes Zr, but OCH₃ does not. The reaction is also sensitive to steric hindrance (Eq. 4.130). Given a choice between an allylic methyl ether and the substantially larger t-butyldimethylsilyl ether, reaction occurs exclusively at the less hindered alkene.

Eq. 4.129

Eq. 4.130

70% yield
90:10 diast.
>99% regio.

The other class of synthetically useful reductive cyclodimerizations of alkenes and alkynes comes from the opposite end of the periodic chart, being catalyzed by Ni(0) d^{10} complexes.[141] The general process is shown in Eq. 4.131. In this case the metallacycle is not stable, and has not been observed, but is inferred from the products.

Eq. 4.131

In the absence of added phosphine, simple cyclodimerization, followed by β-elimination/reductive elimination occurs (Eq. 4.132), giving dienes in reasonable yield. With added tributylphosphine, insertion of an isonitrile (which is isoelectronic and isostructural with CO, and inserts readily into metal carbon σ-bonds) occurs, giving the bicyclic imine which can be hydrolyzed to the ketone (Eq. 4.132). Diynes undergo similar cyclizations to give cyclopentadienone imines.

Eq. 4.132

There are a number of closely related reductive cyclizations of unsaturated species which, however, proceed via π-allylmetal intermediates. These are discussed in Chapter 9.

References

1. Kharasch, M.S.; Tawney, P.O. *J. Am. Chem. Soc.* **1941**, *63*, 2308.
2. For early reviews see: (a) Posner, G.H. *Org. React.* **1972**, *19*, 1. (b) **1979**, *22*, 253. (c) *An Introduction to Synthesis Using Organocopper Reagents*, Wiley, New York, 1980.
3. For a recent review, see: Lipshutz, B.H.; Sengupta, S. *Org. React.* **1992**, *41*, 135.
4. (a) For a review, see: Taylor, J.K. *Synthesis* **1985**, 365. (b) Suzuki, M; Yanagisawa, A.; Noyori, R. *J. Am. Chem. Soc.* **1988**, *110*, 4718.
5. Posner, G.H.; Whitten, C.E.; Sterling, J.J. *J. Am. Chem. Soc.* **1973**, *95*, 7788.
6. (a) Bertz, S.H.; Dabbagh, G.; Villacorta, G.M. *J. Am. Chem. Soc.* **1982**, *104*, 5824. (b) Bertz, S.H.; Dabbagh, G. *J. Org. Chem.* **1984**, *49*, 1119.
7. Martin, S.F.; Fishpaugh, J.R.; Power, J.M.; Giolando, D.M.; Jones, R.A.; Nunn, C.M.; Cowley, A.H. *J. Am. Chem. Soc.* **1988**, *110*, 7226.
8. For reviews see: (a) Lipshutz, B.H.; Wilhelm, R.S.; Kozlowski, T.J. *Tetrahedron* **1984**, *40*, 5005. (b) Lipshutz, B.H. *Syn. Lett.* **1990**, 119. (c) For a different opinion, see: Bertz, S.H. *J. Am. Chem. Soc.* **1990**, *112*, 4031; Snyder, J.P.; Tipsword, G.E.; Spangler, D.P. *J. Am. Chem. Soc.* **1992**, *114*, 1507.
9. (a) Lipshutz, B.H.; Siegmann, K.; Garcia, E. *J. Am. Chem. Soc.* **1991**, *113*, 8161. (b) For a review see: Lipshutz, B.H.; Siegermann, K.; Garcia, E. *Tetrahedron* **1992**, *48*, 2579.
10. Ghribi, A.; Alexakis, A.; Normant, J.F. *Tetrahedron Lett.* **1984**, *25*, 3075.
11. (a) For a review see: Alexakis, A.; Mangeney, P.; Ghribi, A.; Marek, I.; Sedrani, R.; Guir, C.; Normant, J.F. *Pure Appl. Chem.* **1988**, *60*, 49. (b) Mangeney, P.; Alexakis, A.; Normant, J.F. *Tetrahedron Lett.* **1987**, *28*, 2363.

12. (a) For reviews see: Oppolzer, W. *Tetrahedron* **1987**, *43*, 1969, 4057. (b) Rossiter, B.E.; Swingle, N.M. *Chem. Rev.* **1992**, *92*, 771. (c) Oppolzer, W.; Kingma, A.J. *Helv. Chim Acta* **1989**, *72*, 1337.

13. Bergdahl, M.; Nilsson, M.; Olsson, T. *J. Organomet. Chem.* **1990**, *391*, C19.

14. Tanaka, K.; Ushio, H.; Kawabata, Y.; Suzuki, H. *J. Chem. Soc. Perkins I* **1991**, 1445.

15. (a) Rieke, R.D.; Wu, T.C. *Syn. Comm.* **1989**, *19*, 1833. (b) Stack, D.E.; Dawson, B.T.; Rieke, R.D. *J. Am. Chem. Soc.* **1991**, *113*, 4672.

16. Wehmeyer, R.M.; Rieke, R.D. *J. Org. Chem.* **1987**, *52*, 5056, 5057.

17. Zhu, L.; Wehmeyer, R.M.; Rieke, R. *J. Org. Chem.* **1991**, *56*, 1445.

18. Knoess, H.P.; Furlong, M.T.; Rozema, M.J.; Knochel, P. *J. Org. Chem.* **1991**, *56*, 5974.

19. Knochel, P. *J. Am. Chem. Soc.* **1990**, *112*, 7431.

20. Berk, S.C.; Yeh, M.C.P.; Jeong, N.; Knochel, P. *Organometallics* **1990**, *9*, 3053.

21. Majid, T.N.; Knochel, P. *Tetrahedron Lett.* **1990**, *31*, 4413.

22. Rao, S.A.; Tucker, C.E.; Knochel, P. *Tetrahedron Lett.* **1990**, *31*, 7575.

23. Rao, S.A.; Chou, T.S.; Schipor, I.; Knochel, P. *Tetrahedron* **1992**, *48*, 2025.

24. Rozema, M.J.; Sidduri, A.R.; Knochel, P. *J. Org. Chem.* **1992**, *57*, 1956.

25. Tucker, C.E.; Majid, T.N.; Knochel, P. *J. Am. Chem. Soc.* **1992**, *114*, 3983.

26. Linderman, R.J.; Griedel, B.D. *J. Org. Chem.* **1990**, *55*, 5428. (b) Linderman, R.J.; McKenzie, J.R. *J. Organomet. Chem.* **1989**, *361*, 31. (c) Linderman, R.J.; Godfrey, A.J. *J. Am. Chem. Soc.* **1988**, *110*, 6249.

27. (a) Alexakis, A.; Commercon, A.; Coulentianos, C.; Normant, J.F. *Pure Appl. Chem.* **1983**, *55*, 1759. (b) Normant, J.F.; Alexakis, A. *Synthesis* **1981**, 841.

28. Rao, S.A.; Knochel, P. *J. Am. Chem. Soc.* **1991**, *113*, 5735.

29. Trost, B.M. *Acct. Chem. Res.* **1990**, *23*, 34

30. Trost, B.M.; Shi, Y. *J. Am. Chem. Soc.* **1991**, *113*, 701.

31. Trost, B.M.; Shi, Y. *J. Am. Chem. Soc.* **1992**, *114*, 791.

32. Schwartz, J.; Labinger, J.A. *Angew. Chem. Int. Ed.* **1976**, *15*, 333.

33. Gibson, T. *Tetrahedron Lett.* **1982**, *23*, 157.

34. Bertelo, C.A.; Schwartz, J. *J. Am. Chem. Soc.* **1976**, *98*, 262.

35. Loots, M.J.; Schwartz, J. *Tetrahedron Lett.* **1978**, 4381; *J. Am. Chem. Soc.* **1977**, *99*, 8045.

36. Babiak, K.A.; Behling, J.R.; Dygos, J.H.; McLaughlin, K.T.; Ng, J.S.; Kalish, V.J.; Kramer, S.W.; Shone, R.L. *J. Am. Chem. Soc.* **1990**, *112*, 7441.

37. Lipshutz, B.H.; Ellsworth, E.L. *J. Am. Chem. Soc.* **1990**, *112*, 7440.

38. VanHorne, D.E.; Negishi, E-I. *J. Am. Chem. Soc.* **1978**, *100*, 2252.

39. Lipshutz, B.H.; Dimock, S.H. *J. Org. Chem.* **1991**, *56*, 5761.

40. Morris, I.K.; Snow, K.M.; Smith, N.W.; Smith, K.M. *J. Org. Chem.* **1990**, *55*, 1231.

41. Larock, R.C.; Stinn, D.E.; Kuo, M-Y. *Tetrahedron Lett.* **1990**, *31*, 17.

42. Ruth, J.L.; Bergstrom, D.E. *J. Org. Chem.* **1978**, *43*, 2870; Bergstrom, D.E.; Inoue, H.; Reddy, P.A. *J. Org. Chem.* **1982**, *47*, 2174; Hacksell, U.; Daves, G.D., Jr. *J. Org. Chem.* **1983**, *48*, 2870.

43. Somei, M.; Kawesaki, T. *Chem. Pharm. Bull* **1989**, *37*, 3426.

44. Outten, R.A.; Daves, Jr., J.D. *J. Org. Chem.* **1987**, *52*, 5064.

45. Cambie, R.C.; Metzler, M.R.; Rutledge, P.S.; Woodgate, P.D. *J. Organomet. Chem.* **1992**, *429*, 59.

46. Tamao, K.; Sumitani, K.; Kiso, Y.; Zembayashi, M.; Fujioka, H.; Kodama, S-I.; Nakajima, I.; Minato, A.; Kumada, M. *Bull Chem. Soc. Jpn.* **1976**, *49*, 1958.

47. For a review see: Kalinin, V.N. *Synthesis* **1992**, 413.

48. (a) Hayashi, T.; Konishi, M.; Ito, H.; Kumada, M. *J. Am. Chem. Soc.* **1982**, *104*, 4962. (b) Hayashi, T.; Konishi, M.; Fukushima, M.; Kanehira, K.; Hioki, T.; Kumada, M. *J. Org. Chem.* **1983**, *48*, 2195. (c) For a review see: Sawamura, M.; Ito, Y. *Chem. Rev.* **1992**, *92*, 857.
49. (a) Kobayashi, M.; Negishi, E-I. *J. Org. Chem.* **1980**, *45*, 5223. (b) Negishi, E-I.; Owczarczyk, Z. *Tetrahedron Lett.* **1991**, *32*, 6683.
50. Russell, C.E.; Hegedus, L.S. *J. Am. Chem. Soc.* **1983**, *105*, 943.
51. Tius, M.; Gomez-Galeno, J.; Gu, X-Q.; Zaidi, J. *J. Am. Chem. Soc.* **1991**, *113*, 5775.
52. Vincent, P.; Beaucourt, J.P.; Pichat, L. *Tetrahedron Lett.* **1982**, *23*, 63.
53. Okukado, N.; Van Horn, D.E.; Klima, D.L.; Negishi, E-I. *Tetrahedron Lett.* **1978**, 1027.
54. (a) For a review see: Negishi, E-I. *Acct. Chem. Res.* **1982**, *15*, 340. (b) Negishi, E-I.; Okukado, N.; King, A.O.; Van Horn, D.E.; Spiegel, B.I. *J. Am. Chem. Soc.* **1978**, *100*, 2254.
55. Miyaura, N.; Maeda, K.; Suginome, H.; Suzuki, A. *J. Org. Chem.* **1982**, *47*, 2117.
56. Uenishi, J-I.; Beau, J-M.; Armstrong, R.M.; Kishi, Y. *J. Am. Chem. Soc.* **1987**, *109*, 4756.
57. Miyaura, N.; Ishikawa, M.; Suzuki, A. *Tetrahedron Lett.* **1992**, *33*, 2571.
58. For a review see: Snieckus, V. *Chem. Rev.* **1990**, *90*, 879.
59. Sharp, M.J.; Cheng, W.; Snieckus, V. *Tetrahedron Lett.* **1987**, *28*, 5093, 5097.
60. Alves, T.; deOliveira, A.B.; Snieckus, V. *Tetrahedron Lett.* **1988**, *29*, 2135.
61. Muller, D.; Fleury, J-P. *Tetrahedron Lett.* **1991**, *32*, 2229.
62. (a) Fu, J-M.; Sharp, M.J.; Snieckus, V. *Tetrahedron Lett.* **1988**, *29*, 5459, 5463. (b) Siddiqui, M.A.; Snieckus, V. *Tetrahedron Lett.* **1990**, *31*, 1523.
63. For reviews see: (a) Stille, J.K. *Angew. Chem. Int. Ed. Engl.* **1986**, *25*, 508. (b) Mitchell, T.N. *Synthesis* **1992**, 803.
64. (a) Labadie, J.W.; Teuting, D.; Stille, J.K. *J. Org. Chem.* **1983**, *48*, 4634. (b) Labadie, J.W.; Stille, J.K. *J. Am. Chem. Soc.* **1983**, *109*, 669.
65. Kende, A.S.; Roth, B.; Sanfillipo, P.J.; Blacklock, T.J. *J. Am. Chem. Soc.* **1982**, *104*, 5808.
66. Baldwin, J.E.; Adlington, R.M.; Ramcharitar, S.H. *Tetrahedron* **1992**, *48*, 2957.
67. Four, P.; Guibe, F. *J. Org. Chem.* **1981**, *46*, 4439.
68. Stille, J.K.; Simpson, J.H. *J. Am. Chem. Soc.* **1987**, *109*, 2138.
69. Stille, J.K.; Groh, B.L. *J. Am. Chem. Soc.* **1987**, *109*, 813.
70. Evans, D.A.; Black, W.C. *J. Am. Chem. Soc.* **1992**, *114*, 2260.
71. Kalivretenos, A.; Stille, J.K.; Hegedus, L.S. *J. Org. Chem.* **1991**, *56*, 2883.
72. (a) For a review on enol triflates see: Scott, W.J.; McMurray, J.E. *Acc. Chem. Res.* **1988**, *21*, 47. (b) Echavarren, A.M.; Stille, J.K. *J. Am. Chem. Soc.* **1987**, *109*, 5478.
73. Roth, G.P.; Fuller, C.E. *J. Org. Chem.* **1991**, *56*, 3493.
74. Laborde, E.; Lesheski, L.E.; Kiely, J.S. *Tetrahedron Lett.* **1990**, *31*, 1837.
75. Peña, M.R.; Stille, J.K. *J. Am. Chem. Soc.* **1989**, *111*, 5417.
76. Farina, V.; Baker, S.R.; Benigni, D.A.; Hauck, S.I.; Sapino, C., Jr. *J. Org. Chem.* **1990**, *55*, 5833.
77. Saá, J.M.; Martorell, G.; García-Raso, A. *J. Org. Chem.* **1992**, *57*, 678.
78. Gore, W.F.; Wright, M.E.; Davis, P.D.; Labadie, S.S.; Stille, J.K. *J. Am. Chem. Soc.* **1984**, *106*, 6417.
79. Crisp, G.T.; Scott, W.J.; Stille, J.K. *J. Am. Chem. Soc.* **1984**, *106*, 7500.
80. Stille, J.K.; Su, H.; Hill, D.H.; Schneider, P.; Tanaka, M.; Morrison, D.L.; Hegedus, L.S. *Organometallics* **1991**, *10*, 1993.
81. Hiyama, T. *Syn. Lett.* **1991**, 845.
82. Rossi, R.; Carpita, A.; Messeri, T. *Gazz Chem. Ital.* **1992**, *122*, 65.

83. For total syntheses of these compounds utilizing this reaction see: Magnus, P.; Carter, P.; Elliot, J.; Lewis, R.; Harling, J.; Pitterns, T.; Batwa, W.E.; Fort, S. *J. Am. Chem. Soc.* **1992**, *114*, 2544.

84. Magnus, P.; Annoura, H.; Harling, J. *J. Org. Chem.* **1990**, *55*, 1709.

85. Butlin, R.J.; Holmes, A.B.; Mcdonald, E. *Tetrahedron Lett.* **1988**, *29*, 2989.

86. Yang, Z-Y.; Burton, D.B. *Tetrahedron Lett.* **1990**, *31*, 1369.

87. Crombie, L.; Horsham, M.A.; Blade, R.J. *Tetrahedron Lett.* **1987**, *28*, 4879.

88. Torneiro, M.; Fall, Y.; Castedo, L.; Mouriño, A. *Tetrahedron Lett.* **1992**, *33*, 105.

89. Lee, A.S.; Norman, A.W.; Okamura, W.H. *J. Org. Chem.* **1992**, *57*, 3846.

90. Roth, G.P.; Thomas, J.A. *Tetrahedron Lett.* **1992**, *33*, 1959.

91. Rizzo, C.J.; Smith, A.B., Jr. *Tetrahedron Lett.* **1988**, *29*, 2793.

92. Thompson, S.K.; Heathcock, C.H. *J. Org. Chem.* **1990**, *55*, 3004.

93. (a) Cowell, A.; Stille, J.K. *J. Am. Chem. Soc.* **1980**, *102*, 4193. (b) Martin, L.D.; Stille, J.K. *J. Org. Chem.* **1982**, *47*, 3630.

94. Mori, M.; Chiba, K.; Ban, Y. *J. Org. Chem.* **1978**, *43*, 1684.

95. (a) Mori, M.; Chiba, K.; Okita, K.; Ban, Y. *Chem. Comm.* **1979**, 698. (b) *Tetrahedron* **1985**, *41*, 387.

96. Tilley, J.W.; Coffen, D.L.; Schaer, B.H.; Lind, J. *J. Org. Chem.* **1987**, *52*, 2469.

97. Meyers, A.I.; Robichaud, A.J.; McKennon, M.J. *Tetrahedron Lett.* **1992**, *33*, 1181.

98. Perry, R.J.; Wilson, B.D.; Miller, R.J. *J. Org. Chem.* **1992**, *57*, 2883.

99. Shimoyama, I.; Zang, Y.; Wu, G.; Negishi, E-I. *Tetrahedron Lett.* **1990**, *31*, 2841.

100. For a review see: Heck, R.F. *Org. React.* **1982**, *27*, 345. R.F. Heck is a giant in organometallic chemistry, having pioneered the development of synthetically useful organometallic reactions which span the entire transition series.

101. McCrindle, R.; Ferguson, G.; Arsenault, G.J.; McAlees, A.J.; Stephenson, D.K. *J. Chem. Res. Synop.* **1984**, 360.

102. Andersson, C.M.; Karabelas, K.; Hallberg, A.; Andersson, C. *J. Org. Chem.* **1985**, *50*, 3891.

103. (a) Jeffry, T. *Tetrahedron Lett.* **1985**, *26*, 2667. (b) Larock, R.C.; Baker, B.E. *Tetrahedron Lett.* **1988**, *29*, 905. (c) Jeffrey, T. *Chem. Comm.* **1984**, 1287. (d) Jeffry, T. *Synthesis* **1987**, 70.

104. Amatore, C.; Azzabi, M.; Jutand, A. *J. Am. Chem. Soc.* **1991**, *113*, 8375.

105. Spencer, A. *J. Organomet. Chem.* **1983**, *258*, 101; **1984**, *265*, 323; **1984**, *270*, 115.

106. Daves, G.D., Jr.; Hallberg, A. *Chem. Rev.* **1989**, *89*, 1433.

107. (a) Cabri, W.; Candiani, I.; Bedeschi, A.; Penco, S. *J. Org. Chem.* **1992**, *57*, 1481. (b) Cabri, W.; Candiani, I.; Bedeschi, A.; Santi, R. *J. Org. Chem.* **1992**, *57*, 3558.

108. Hirota, K.; Kitade, Y.; Isobe, Y.; Maki, Y. *Heterocycles* **1987**, *26*, 355.

109. Harrington, P.J.; Hegedus, L.S. *J. Org. Chem.* **1984**, *49*, 2657.

110. Farr, R.N.; Outten, R.A.; Chen, J.C-Y.; Daves, Jr., G.D. *Organometallics* **1990**, *9*, 3151.

111. Sundberg, R.J.; Cherney, R.J. *J. Org. Chem.* **1990**, *55*, 6028.

112. Liang, S.; Paquette, L.A. *Acta Chem. Scand.* **1992**, *46*, 597.

113. Abelman, M.M.; Oh, T.; Overman, L.E. *J. Org. Chem.* **1987**, *52*, 4130.

114. (a) Ozawa, F.; Kubo, A.; Hayashi, T. *Tetrahedron Lett.* **1992**, *33*, 1485. (b) *J. Organomet. Chem.* **1992**, *428*, 267.

115. Sato, Y.; Watanabe, S.; Shibasaki, M. *Tetrahedron Lett.* **1992**, *33*, 2589, 2593.

116. Negishi, E-I.; Sawada, H.; Tour, J.M.; Wei, Y. *J. Org. Chem.* **1988**, *53*, 913.

117. Zhang, Y.; Negishi, E-I. *J. Am. Chem. Soc.* **1989**, *111*, 3454.
118. Abelman, M.M.; Overman, L.E. *J. Am. Chem. Soc.* **1988**, *110*, 2328.
119. Zhang, Y.; Wu, G.; Agnel, G.; Negishi, E-I. *J. Am. Chem. Soc.* **1990**, *112*, 8590.
120. Meyer, F.E.; Parsons, P.J.; deMeijere, A. *J. Org. Chem.* **1991**, *56*, 6487.
121. (a) Torii, S.; Okumoto, H.; Nishimura, A. *Tetrahedron Lett.* **1991**, *32*, 4167. (b) Negishi, E-I.; Harring, L.S.; Owczarczyk, Z.; Mohamad, M.M.; Ay, M. *Tetrahedron Lett.* **1992**, *33*, 3253.
122. Trost, B.M.; Pfrengle, W.; Urabe, H.; Dumas, J. *J. Am. Chem. Soc.* **1992**, *114*, 1923.
123. Nuss, J.M.; Levine, B.H.; Rennels, R.A.; Heravi, M.M. *Tetrahedron Lett.* **1991**, *32*, 5243.
124. Arcadi, A.; Burini, A.; Cacchi, S.; Delmastro, M.; Marinelli, F.; Pietroni, B.R. *J. Org. Chem.* **1992**, *57*, 976.
125. For reviews see: (a) Ryabov, A.D. *Synthesis* **1985**, 233. (b) Omae, I. *Chem. Rev.* **1979**, *79*, 287. (c) Pfeffer, M.; Dehand, P. *Coord. Chem. Rev.* **1976**, *18*, 327.
126. Brisdon, B.J.; Nair, P.; Dyke, S.F. *Tetrahedron* **1981**, *37*, 173.
127. Horino, H.; Inoue, N. *J. Org. Chem.* **1981**, *46*, 4416.
128. Carr, K.; Sutherland, J.K. *J. Chem. Soc., Chem. Comm.* **1984**, 1227.
129. Swanson, D.R.; Rousset, C.J.; Negishi, E-I.; Takahashi, T.; Seki, T.; Saburi, M.; Uchida, Y. *J. Org. Chem.* **1989**, *54*, 3521.
130. Fagan, P.J.; Nugent, W.A. *J. Am. Chem. Soc.* **1988**, *110*, 2310.
131. (a) Nugent, W.A.; Taber, D.F. *J. Am. Chem. Soc.* **1989**, *111*, 6435. (b) Negishi, E-I.; Holmes, S.J.; Tour, J.M.; Miller, J.A.; Cederbaum, F.E.; Swanson, D.R.; Takahashi, T. *J. Am. Chem. Soc.* **1989**, *111*, 3336. (c) Negishi, E-I.; Miller, S.R. *J. Org. Chem.* **1989**, *54*, 6014.
132. Rajanbabu, T.V.; Nugent, W.A.; Taber, D.F.; Fagan, P.J. *J. Am. Chem. Soc.* **1988**, *110*, 7128.
133. (a) Agnel, G.; Negishi, E-I. *J. Am. Chem. Soc.* **1991**, *113*, 7424. (b) Agnel, G.; Owczarczyk, Z.; Negishi, E-I. *Tetrahedron Lett.* **1992**, *33*, 1543.
134. Mori, M.; Uesaka, N.; Shibasaki, M. *J. Org. Chem.* **1992**, *57*, 3519.
135. (a) Buchwald, S.L.; Watson, B.T.; Wannamaker, M.W.; Dewan, J.C. *J. Am. Chem. Soc.* **1989**, *111*, 4486. (b) Coles, N.; Whitby, R.J.; Blagg, J. *Syn. Lett.* **1992**, 143.
136. Buchwald, S.L.; Wannamaker, M.W.; Watson, B.T. *J. Am. Chem. Soc.* **1989**, *111*, 776.
137. Jensen, M.; Livinghouse, T. *J. Am. Chem. Soc.* **1989**, *111*, 4495.
138. Knight, K.S.; Waymouth, R.M. *J. Am. Chem. Soc.* **1991**, *113*, 6268.
139. Takahashi, T.; Seki, T.; Nitto, Y.; Saburi, M.; Rousset, C.J.; Negishi, E-I. *J. Am. Chem. Soc.* **1991**, *113*, 6266.
140. (a) Hoveyda, A.H.; Xu, Z. *J. Am. Chem. Soc.* **1991**, *113*, 5079. (b) Hoveyda, A.H.; Xu, Z,; Morken, J.P.; Houri, A.F. *J. Am. Chem. Soc.* **1991**, *113*, 8950. (c) Hoveyda, A.H.; Morken, J.P.; Houri, A.F.; Xu, Z. *J. Am. Chem. Soc.* **1992**, *114*, 6692.
141. For a review see: Tamao, K.; Kobayashi, K.; Ito, Y. *Syn. Lett.* **1992**, 539.

CHAPTER 5

Synthetic Applications of Transition Metal Carbonyl Complexes

5.1 Introduction

Carbon monoxide coordinates to virtually all transition metals, and an enormous number of complexes with a bewildering array of structures are known. However, most synthetically useful reactions of metal carbonyl complexes, those in which the CO ligand is more than a spectator ligand, involve the first row, homoleptic ("all CO") carbonyl complexes shown in Figure 5.1. Carbon monoxide is among the best π-acceptor ligands, stabilizing low oxidation states and electron density on the metal. Many neutral M° carbonyl complexes can be reduced to very low *formal* oxidation states (-I through -III), and the resulting anions are powerful nucleophiles. The ease of migratory insertion of CO into metal-carbon σ-bonds makes metal carbonyls attractive catalysts for the introduction of carbonyl groups into organic substrates, and the reversibility of this process is central to *decarbonylation* processes. The general reactions that metal carbonyls undergo are shown in Figure 5.2. Specific examples will be discussed in the following sections.

Figure 5.1   Homoleptic Metal Carbonyls

Figure 5.2   Reactions of Metal Carbonyls

5.2 Coupling Reactions of Metal Carbonyls

Allylic halides undergo clean coupling at the less substituted ter-minus when treated with nickel carbonyl in polar solvents such as DMF or THF (Eq. 5.1).[1] The reaction is reasonably well-understood, and involves dissociation of one CO from the coordinatively saturated, very labile,

Eq. 5.1

nickel carbonyl (a) followed by oxidative addition of the allylic halide to the unsaturated Ni(CO)$_3$ fragment (b) giving ultimately the η^3-allyl-nickel carbonyl intermediate (c). This complex has been detected spectro-scopically (ν$_{CO}$ 2060 cm^{-1}) and can be generated independently from CO treatment of η^3-allylnickel bromide dimer. When the reaction is run in apolar solvents such as benzene, coupling does not occur; rather the η^3-allylnickel bromide dimer can be isolated in excellent yield (d) (Chapter 9). However, in DMF both the dimer and the η^3-allylnickel carbonyl bro-mide monomer react quickly with excess allylic bromide to give the coupled product, biallyl. All steps save the last one appear to be re-versible, since simple treatment of η^3-allylnickel bromide dimer with car-bon monoxide in DMF results in efficient coupling, and both allyl bromide and Ni(CO)$_4$ can be detected in solution. Under high dilution conditions α,ω-bis allyl halides are cyclized (Eq. 5.2).[2] In this way simple 12, 14, and 18 membered rings,[3] humulene[4] and macrocyclic lactones[5] have been prepared. This potentially useful coupling has been utilized very little, primarily because nickel carbonyl is a volatile (BP 43°C) highly toxic colorless liquid that is difficult to handle and to dispose.

Eq. 5.2

In contrast, iron pentacarbonyl is considerably less toxic, less volatile, and easier to handle, and a number of synthetically useful procedures involving it have been developed. It is, however, not very labile, and since it is coordinatively saturated, one CO must be lost before reactions can take place. As a consequence, reactions involving $Fe(CO)_5$ frequently require heating, sonication, or photolysis to generate the coordinatively unsaturated $Fe(CO)_4$ fragment. The species $Fe(CO)_4$ can be generated under much milder conditions from the gold/orange dimer, $Fe_2(CO)_9$, prepared by irradiation of $Fe(CO)_5$ in acetic acid. Very gentle heating of solutions (suspensions) of $Fe_2(CO)_9$ releases $Fe(CO)_4$ for reactions in which substrates or products are thermally or photochemically unstable.

Reaction of α,α'-disubstituted-α,α'-dibromoketones with $Fe(CO)_4$ generates an iron(II) oxallyl cation (Eq. 5.3) which undergoes a variety of synthetically interesting [3+4] cycloadditions to give seven-membered rings (Eq. 5.4).[6] Generation of the iron-oxallyl cation is usually done *in situ*, and is likely to involve oxidative addition (nucleophilic attack) of the bromoketone to the electron rich iron(0) species, to give an iron(II) C-enolate. Rearrangement to the O enolate followed by loss of Br^- gives the reactive oxallyl cation. A variety of dienes undergo cycloaddition, including five membered aromatic heterocycles, to give bridge bicyclic systems (Eq. 5.5). α-Substitution on the bromoketone is required to stabilize positive charge in the complex, so dibromoacetone itself does not react.

Eq. 5.3

Eq. 5.4

Eq. 5.5

$X = CH_2, O, NAc$

However, tetrabromoacetone is reactive, permitting the introduction of the unsubstituted acetone fragment, after a subsequent reductive debromination (Eq. 5.6).[7] This strategy has been used to synthesize tropane alkaloids (Eq. 5.7).[8] Cycloheptatrienone undergoes cycloaddition to the carbonyl group instead of the olefin, probably because of the highly stabilized cation resulting from initial attack at oxygen (Eq. 5.8).[9]

Iron oxallyl cation complexes also undergo cycloaddition reactions with electron-rich alkenes, such as styrene or enamines, to produce cyclopentanones (Eq. 5.9).[10] Even relatively nonnucleophilic olefins such as N-tosyl enamines react cleanly (Eq. 5.10).[11]

Eq. 5.6

Eq. 5.7

Eq. 5.8

Eq. 5.9

Eq. 5.10

5.3   Carbonylation Reactions[12a]

By far the most common synthetic reactions of transition metal carbonyls are those which introduce a carbonyl group into the product. A large number of industrial processes, including hydroformylation (oxo), hydrocarboxylation, and the Monsanto acetic acid process, are predicated on the facility with which CO inserts into metal carbon bonds. Carbonylation reactions are also useful in the synthesis of fine chemicals.

Although nickel carbonyl couples allylic halides in polar solvents, in less polar solvent in the presence of methanol they are converted to β,γ-unsaturated esters (Eq. 5.11). Under these conditions CO insertion/ methanol cleavage is faster than reaction with allyl halides (coupling) which requires very polar solvents.

Eq. 5.11

Iron carbonyl promotes a very unusual cyclocarbonylation of allyl epoxides to produce β-lactones or γ-lactones, depending on conditions.[12b,c] The process is somewhat complex, and involves treatment of vinyl epoxides with $Fe(CO)_4$, generated from $Fe(CO)_5$ by heat, sonication or irradiation, or from $Fe_2(CO)_9$ under milder conditions (Eq. 5.12). $Fe(CO)_4$ is a good nucleophile, and can attack (oxidative addition) the vinyl epoxide in an S_N2' manner to give a cationic σ-alkyliron(II) carbonyl complex (a). Cationic metal carbonyl complexes are prone to nucleophilic attack at a carbonyl group, in this case producing a "ferrilactone" complex (b), best represented as the η³-allyl complex, which is stable and isolable. Oxidation of this complex with cerium(IV) leads to the β-lactone (c), presumably by an "oxidatively driven reductive elimination" from the five membered η¹-allyl complex. In contrast, decomposition of the η³-allyl complex by treatment with high pressures of carbon monoxide at elevated temperatures produces the δ-lactone, formally by reductive elimination from the *seven* membered η¹-allyl complex. This chemistry has been used in the total synthesis of a variety of natural products (Eq. 5.13[12] and Eq. 5.14[13]).

Eq. 5.12

Eq. 5.13

Eq. 5.14

These η^3-allyliron lactone complexes can also be used to synthesize β-lactams. In the presence of Lewis acids, amines will attack the allyl complex in an S_N2' manner to produce a rearranged η^3-allyliron lactam complex. Oxidation of this complex produces β-lactams (Eq. 5.15).[14] The same η^3-allyliron lactam complexes are available directly from the amino alcohol, and it is likely that the conversion of the lactone complex to the lactam complex proceeds through the amino alcohol.

In the above carbonylations, the first step is nucleophilic attack of the zero valent metal on the substrate (S_N2-like oxidative addition, Chapter 2). Anionic metal carbonyl complexes are even stronger nucleophiles, and affect a number of synthetically useful carbonylation reactions.

Eq. 5.15

Probably the most intensively studied and highly developed (for organic synthesis) anionic metal carbonyl species is disodium tetracarbonylferrate, $Na_2Fe(CO)_4$, Collman's reagent.[15] This d^{10} Fe(-II) complex is easily prepared by reduction of the commercially available iron pentacarbonyl by sodium-benzophenone ketyl in dioxane at reflux. (This complex is also available commercially.)

This complex is useful in synthesis specifically because of its high nucleophilicity and the ease of the CO-insertion reaction in this system. Figure 5.3 summarizes this chemistry. Organic halides and tosylates react with $Na_2Fe(CO)_4$ with typical S_N2 kinetics (second-order), stereochemistry (inversion), and order of reactivity (CH_3 > RCH_2 > R'RCH; RI > RBr > ROTs > RCl; vinyl, aryl, inert) to produce coordinatively saturated anionic d^8 alkyliron(0) complexes (a). These undergo protolytic cleavage to produce the corresponding hydrocarbon, an overall reduction of the halide (b). In the presence of excess carbon monoxide or added triphenylphosphine, migratory insertion takes place to form the acyl complex (c), accessible directly by the reaction of $Na_2Fe(CO)_4$ with acid halides (d). This acyl complex reacts with acetic acid to produce aldehydes, providing a high-yield conversion of alkyl and acyl halides to aldehydes (e). Oxidative cleavage of either the acyl or alkyliron(0) complex by oxygen (f) or halogen (g) produces carboxylic acid derivatives. Finally, the acyliron(0) complex itself is still sufficiently nucleophilic to react with reactive alkyl iodides to give unsymmetric ketones in excellent yield (h). Interestingly, the alkyliron(0) complex reacts in a similar fashion with alkyl iodides to produce unsymmetric ketones inserting CO along the way (h). Hence this chemistry provides routes from alkyl and acid halides to alkanes, aldehydes, ketones, and carboxylic acid derivatives.

The reagent is quite specific for halides. Ester, ketone, nitrile, and olefin functionality is tolerated elsewhere in the molecule. In mixed halides (i.e., chloro-bromo compounds) the more reactive halide is the exclusive site of reactivity. The major limitation is the high basicity (pK_b about that of OH^-) of $Na_2Fe(CO)_4$, which results in competing elimination reactions with tertiary and secondary substrates. Additionally, allylic halides having alkyl groups δ to the halide fail as substrates, since stable 1,3-diene iron complexes form preferentially. Finally, since the migratory insertion fails when the R group contains adjacent electronegative groups, such as halogen or alkoxy groups, the syntheses involving insertion are limited to simple primary or secondary substrates.

The chemistry described above has been subjected to very close experimental scrutiny, and as a consequence, the mechanistic features are understood in detail. Both the alkyliron(0) complex and the acyliron(0) complex have been isolated as air-stable, crystalline $[(Ph_3P)_2N]^+$ salts, fully characterized by elemental analysis and IR and NMR spectra. These salts undergo the individual reactions detailed in Figure 5.3. Careful kinetic studies indicate that ion-pairing effects dominate the reactions of $Na_2Fe(CO)_4$, and account for the observed 2×10^4 rate increase (tight ion pair $[NaFe(CO)_4]^-$ versus solvent-separated ion pair $[Na^+:S:Fe(CO)_4]^-$ as

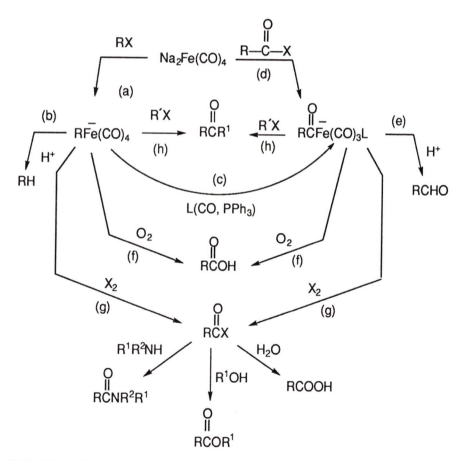

Figure 5.3   Reaction Chemistry of $Na_2Fe(CO)_4$

the kinetically active species) in going from THF to *N*-methylpyrroli-done. The rate law, substrate order of reactivity, stereochemistry at carbon, and activation parameters (particularly the large negative entropy of activation) are all consistent with an S_N2-type oxidative-addition, with no competing one-electron mechanism. The migratory-insertion step is also subject to ion-pairing effects, and is accelerated by tight ion pairing $[Li^+ > Na^+ > Na\text{-crown}^+ > (Ph_3P)_2N^+]$. The insertion reaction is overall second-order, first-order in both $NaRFe(CO)_4$ and added ligand, with about a twenty-fold difference in rate over the range of ligands Ph_3P (slowest) to Me_3P (fastest).[16]

An interesting variation of this chemistry[17] results in the production of ethyl ketones from organic halides, $Na_2Fe(CO)_4$, and ethylene (Eq. 5.16). The reaction is thought to proceed via the acyl iron(0) complex, which inserts ethylene into the acyl-metal bond and rapidly rearranges to the α-metalloketone, which then cannot further insert ethylene since the R group now contains an adjacent electronegative acyl group. Intramolecular versions of this can be used to synthesize cyclic ketones (Eq. 5.17),[18] but in some cases is not regioselective (Eq. 5.18).[19]

Eq. 5.16

Eq. 5.17

Eq. 5.18

1:1, 55%

Anionic nickel acyl complexes, generated *in situ* by the reaction of organolithium reagents with the toxic, volatile nickel carbonyl, affect a number of synthetically useful transformations including the acylation of allylic halides, 1,4-acylation of conjugated enones, and 1,2-addition to quinones (Eq. 5.19).[20] Because of the toxicity of the reagent, few groups have had the courage to use it, although one nice synthetic application has been reported (Eq. 5.20).[21] The same general reaction chemistry can be achieved by substituting for nickel carbonyl the orange, air stable, non-volatile crystalline solid $(PPh_3)(CO)_2(NO)Co$, a complex much more easily handled.[22] No one has yet seen fit to use this in synthesis, however. (This complex is isoelectronic with $Ni(CO)_4$, a nickel(0), d^{10} complex, since the nitrosyl ligand is formally a cation, making this a cobalt (-1), d^{10} complex.)

Eq. 5.19

Vinyl and aryl iodides are converted to esters by treatment with nickel carbonyl and sodium methoxide in methanol (Eq. 5.21).[23] The mechanism of this process is not known but is likely to involve an anionic nickel acyl species. An intramolecular version has also been developed (Eq. 5.22).[23]

Eq. 5.20

Eq. 5.21

Eq. 5.22

5.4 Decarbonylation Reactions[24]

Much of this chapter has been devoted to consideration of methods for the introduction of carbonyl groups into organic substrates via a "migratory insertion" reaction, in which a metal alkyl or aryl group migrates to an adjacent, coordinated CO to produce a σ-acyl complex, which is then cleaved to produce the organic carbonyl compound. The reverse process, in which organic carbonyl compounds (specifically aldehydes and acid chlorides) are decarbonylated, is also possible, and often quite useful in organic synthesis. Although a number of transition-metal complexes will function as decarbonylation agents, by far the most efficient is $(Ph_3P)_3RhCl$, the Wilkinson hydrogenation catalyst. This complex reacts with alkyl, aryl, and vinyl aldehydes under mild conditions to produce the corresponding hydrocarbon and the very stable $(Ph_3P)_2Rh(Cl)CO$, as in Eq. 5.23. The course of this stoichiometric decarbonylation can easily be monitored by the change in color of the solution, from the deep red of $(Ph_3P)_3RhCl$ to the canary yellow of $(Ph_3P)_3Rh(Cl)CO$. *trans*-α-Alkylcinnamaldehydes are decarbonylated with retention of olefin geometry, producing *cis*-substituted styrenes.[25] More impressively, chiral aldehydes are converted to chiral hydrocarbons with overall retention and with a high degree of stereoselectivity (Eq. 5.24).[26] The mechanism proposed for this

Eq. 5.23

Eq. 5.24

Eq. 5.25

decarbonylation is that in Eq. 5.25. It involves, as a first step, oxidative addition of the aldehyde to the rhodium complex. The next step is the reverse of the carbonyl insertion reaction, that is, migration of the alkyl group from carbonyl to metal. It is well-known that migratory insertion is a reversible process, and can be driven in either direction by appropriate choice of reaction conditions. In addition, it has been shown in other cases that this reversible transformation proceeds with retention of stereochemistry of the migrating alkyl group; hence the retention of stereochemistry in the decarbonylation of chiral aldehydes. The last step is the reductive elimination of alkane (RH) and the production of $RhCl(CO)L_2$. This last step is irreversible (alkanes do not oxidatively add to $RhCl(CO)L_2$), and drives the entire process to completion. Since $RhCl(CO)L_2$ is much less reactive in oxidative addition reactions than $RhClL_3$ (CO is a π acceptor and withdraws electron density from the metal), it does not react with aldehydes under these mild conditions, and the decarbonylation described above is stoichiometric. When the reaction is carried out at temperatures in excess of 200°, both $RhClL_3$ and $RhCl(CO)L_2$ function as decarbonylation catalysts, presumably because $RhCl(CO)L_2$ will oxidatively add aldehydes under these severe conditions.

Quite complex aldehydes can be decarbonylated without problems (Eq. 5.26[27] and Eq. 5.27[28]).

Eq. 5.26

Eq. 5.27

(*via* β-elimination)

Acid chlorides also undergo decarbonylation upon treatment with RhCl(PPh₃)₃, but the reaction suffers several complications not encountered with aldehydes. The reaction of acid chlorides is most straightforward with substrates that have no β-hydrogens. In these cases, decarbonylation occurs smoothly to give the alkyl chloride. In contrast to aldehydes, the initial oxidative adduct is quite stable, and has been isolated and fully characterized in several instances. Heating this adduct leads to decarbonylation, via a mechanism thought to be strictly analogous to that involved in aldehyde decarbonylation. However, the decarbonylation of acid halides differs from that of aldehydes in several respects. Decarbonylation of optically active acid chlorides with RhCl(PPh₃)₃ produces racemic alkyl chlorides. This is in direct contrast to aldehydes, which undergo decarbonylation with a high degree of retention.

Acid halides with β-hydrogens undergo decarbonylation by RhCl(PPh₃)₃ to produce primarily olefins resulting from β-hydride elimination from the σ-alkyl intermediate, rather than reductive elimination (Eq. 5.28). Branched acid chlorides that can undergo β-elimination in several directions give mixtures of products with the most substituted olefins predominating.

Eq. 5.28

Contrary to older literature reports, aroyl chlorides and α,β-unsaturated acid chlorides do *not* undergo decarbonylation to give the corresponding unsaturated chloride, but rather produce stable aryl metal complexes or quaternary phosphonium salts.[29]

5.5   Metal Acyl Enolates

Many of the metal-acyl species discussed above are quite stable, and can be isolated and handled easily. One complex, CpFe(CO)(PPh₃)(COCH₃), has been developed into a useful synthetic reagent. This complex has several interesting characteristics: (1) it is easy to prepare and handle, (2) it is chiral at iron, and can be resolved, and (3) the protons α to the acyl group are acidic and the corresponding metal acyl enolate undergoes reaction with a variety of electrophiles (Eq. 5.29). More interesting, because the complex is chiral at iron and one face is hindered by the triphenylphosphine ligand, reaction of these acyliron enolates occur with very high stereoselectivity (Eq. 5.30).[30] The observed absolute stereochemistry results from alkylation of the E enolate with the carbonyl group and O⁻ *anti*, from the less hindered face. Oxidative cleavage of the resulting complex produces the carboxylic acid derivative in good yield with high ee.[31]

This chemistry has been used extensively in the synthesis of optically active compounds. Two such cases are shown in Eq. 5.31[32] and Eq. 5.32.[33]

Eq. 5.29

Eq. 5.30

(R)-(−)

(E-enolate, O⁻ *anti* to CO)

E⁺
(unhindered face)

high ee

Eq. 5.31

(R)

(1) BuLi

(2) *t*BuSCH₂Br

[Fe]

71%
(R,S)

(1) Br₂

(2)

CF₃COOH

Hg(OAc)₂

(S,S) captopril

Eq. 5.32

(S)

(1) BuLi
(2) Et₂AlCl

(3)

(1) TsOH

(2) Br₂

45%

As might be expected, α,β-unsaturated iron acyl complexes also undergo reactions in a highly diastereoselective manner. Both conjugate ad-
dition/enolate trapping (Eq. 5.33)[34] and γ-deprotonation-α-alkylation (Eq.
5.34)[35] enjoy a high degree of stereocontrol. Vinylogous iron acyl complexes also undergo a variety of alkylation reactions (Eq. 5.35)[36] although
these have not yet been used in total synthesis.

Eq. 5.33

Eq. 5.34

γ-deprotonation   α-alkylation

Eq. 5.35

50-80%

EX = MeI, Et₃I, BuI, nC₅OTf, ⟍⟋Br , BnBr, ⟍⟍CHO

Other metal acyl enolate complexes, particularly those of molybdenum and tungsten, have been prepared and their reaction chemistry studied (for example Eq. 5.36[37]). However the focus of these studies has been more organometallic/mechanistic than synthetic, so the potential of these systems in the synthesis of complex molecules is yet to be realized.

Eq. 5.36

References

1. Corey, E.J.; Semmelhack, M.F.; Hegedus, L.S. *J. Am. Chem. Soc.* **1968**, *90*, 2416.
2. Dauben, W.G.; Beasley, G.H.; Broadhurst, M.D.; Muller, B.; Peppard, D.J.; Pesnelle, P.; Suter, C. *J. Am. Chem. Soc.* **1974**, *96*, 4724.
3. Corey, E.J.; Wat, E. *J. Am. Chem. Soc.* **1967**, *89*, 2757.
4. Corey, E.J.; Hamanaka, E. *J. Am. Chem. Soc.* **1967**, *89*, 2758.
5. Corey, E.J.; Kirst, H.A. *J. Am. Chem. Soc.* **1972**, *94*, 667.
6. (a) For a review, see: Noyori, R. *Acct. Chem. Res.* **1979**, *12*, 61. (b) Takaya, H.; Makino, S.; Hayakawa, Y.; Noyori, R. *J. Am. Chem. Soc.* **1978**, *100*, 1765, 1778.
7. Noyori, R.; Sato, T.; Hayakawa, Y. *J. Am. Chem. Soc.* **1978**, *100*, 2561.
8. Hayakawa, Y.; Baba, Y.; Makino, S.; Noyori, R. *J. Am. Chem. Soc.* **1978**, *100*, 1786.
9. Ishizu, T.; Harano, K.; Yasuda, M.; Kanematsu, K. *J. Org. Chem.* **1981**, *46*, 3630.
10. Hayakawa, Y.; Yokoyama, K.; Noyori, R. *J. Am. Chem. Soc.* **1978**, *100*, 1791, 1799.
11. Hegedus, L.S.; Holden, M. *J. Org. Chem.* **1985**, *50*, 3920.
12. (a) Colquhoun, H.M.; Thompson, D.J.; Twigg, M.V. "Carbonylation", Plenum Press, New York, 1991. (b) For a review see: Ley, S.V. *Phil. Trans R. Soc. Lond. A* **1988**, *326*, 633. (c)Bates, R.W.; Fernandez-Moro, R.; Ley, S.V. *Tetrahedron Lett.* **1991**, *32*, 2651.
13. Kotecha, N.R.; Ley, S.V.; Mantegani, S. *Synlett* **1992**, 395.
14. (a) Annis, G.D.; Hebblethwaite, E.M.; Hodgson, S.T.; Hollingshead, D.M.; Ley, S.V. *J. Chem. Soc., Perkin I* **1983**, 2851. (b) Horton, A.M.; Hollinshead, D.M.; Ley, S.V. *Tetrahedron* **1984**, *40*, 1737.
15. Collman, J.P. *Acct. Chem. Res.* **1975**, *8*, 342.
16. Collman, J.P.; Finke, R.G.; Cawse, J.N.; Brauman, J.I. *J. Am. Chem. Soc.* **1977**, *99*, 2515; *J. Am. Chem. Soc.* **1978**, *100*, 4766.
17. Cooke, M.P., Jr.; Parlman, R.M. *J. Am. Chem. Soc.* **1975**, *97*, 6863.
18. Merour, J.Y.; Roustan, J.L.; Charrier, C.; Collin, J.; Benaim, J. *J. Organomet. Chem.* **1973**, *51*, C24.
19. (a) McMurry, J.E.; Andrus, A.; Ksander, G.M.; Musser, J.H.; Johnson, M.A. *J. Am. Chem. Soc.* **1979**, *101*, 1330. (b) *Tetrahedron* **1981**, *27 Supplement*, 319.
20. Corey, E.J.; Hegedus, L.S. *J. Am. Chem. Soc.* **1969**, *91*, 4926.
21. Semmelhack, M.F.; Keller, L.; Sato, T.; Spiess, E.J.; Wulff, W. *J. Org. Chem.* **1985**, *50*, 5566.
22. Hegedus, L.S.; Perry, R.J. *J. Org. Chem.* **1985**, *50*, 4955.
23. (a) Corey, E.J.; Hegedus, L.S. *J. Am. Chem. Soc.* **1969**, *91*, 1233. (b) Corey, E.J.; Kirst, H.A.; Katzenellenbogen, J.A. *J. Am. Chem. Soc.* **1970**, *92*, 6314.
23. Semmelhack, M.F.; Brickner, S.J. *J. Org. Chem.* **1981**, *46*, 1723.
24. For a review see: Tsuji, J. in *Organic Syntheses via Metal Carbonyls* vol. 2, Wender, I.; Pino, P., eds., Wiley, New York, 1977, pp. 595-654. Catalytic decarbonylation of aldehydes has recently been achieved using Wilkinsons catalyst and $(PhO)_2P(O)N_3$: O'Connor, J.; Ma, J. *J. Org. Chem.* **1992**, *57*, 5075.
25. Tsuji, J.; Ohno, K. *Tetrahedron Lett.* **1967**, 2173.
26. (a) Walborsky, H.M.; Allen, L.E. *Tetrahedron Lett.* **1970**, 823. (b) Walborsky, A.M.; Allen, L.E. *J. Am. Chem. Soc.* **1971**, *93*, 5465.
27. Iley, D.E.; Fraser-Reid, B. *J. Am. Chem. Soc.* **1975**, *97*, 2563.
28. Sargent, M.; Mongrain, M.; Deslonchamps *Can. J. Chem.* **1972**, *50*, 336.

29. (a) Kampmeier, J.A.; Rodehorst, R.M.; Philip, Jr., J.B. *J. Am. Chem. Soc.* **1981**, *103*, 1847. (b) Kampmeier, J.A.; Mahalingam, S. *Organometallics* **1984**, *3*, 489. (c) Kampmeier, J.A.; Harris, S.H.; Rodehorst, R.M. *J. Am. Chem. Soc.* **1981**, *103*, 1478.

30. For reviews see: (a) Davies, S.G. *Pure Appl. Chem.* **1988**, *60*, 13. (b) Blacksburn, B.K.; Davies, S.G.; Sutton, K.H.; Whittaker, M. *Chem. Soc. Rev.* **1988**, *17*, 147.

31. For a review, see: Davies, S.G.; Bashiardes, G.; Beckett, R.P.; Coote, S.J.; Dordor-Hedgecock, I.M.; Goodfellow, C.L.; Gravatt, G.L.; McNally, J.P.; Whittaker, M. *Phil. Trans R. Soc. Lond. A* **1988**, *326*, 619.

32. Bashiardes, G.; Davies, S.G. *Tetrahedron Lett.* **1987**, *28*, 5563.

33. Beckett, R.P.; Davies, S.G.; Mortlock, A.A. *Tetrahedron Asymmetry* **1992**, *3*, 123.

34. Davies, S.G.; Dordor-Hedgecock, I.M.; Easton, R.J.C.; Preston, S.C.; Sutton, K.H.; Walker, J.C. *Bull Soc. Chim. Fr.* **1987**, 608.

35. Davies, S.G.; Easton, R.J.C.; Gonzalez, A.; Preston, S.C.; Sutton, K.H.; Walker, J.C. *Tetrahedron* **1986**, *42*, 3987.

36. Mattson, M.N.; Helquist, P. *Organometallics* **1992**, *11*, 4.

37. Rusik, C.A.; Collins, M.A.; Gamble, A.S.; Tonker, T.L.; Templeton, J.L. *J. Am. Chem. Soc.* **1989**, *111*, 2550.

CHAPTER 6

Synthetic Applications of Transition Metal Carbene Complexes

6.1 Introduction

Carbene complexes — complexes having formal metal-to-carbon double bonds — are known for metals across the entire transition series, although relatively few have been developed as useful reagents for organic synthesis. The two bonding extremes for carbene complexes are represented by the *electrophilic*, heteroatom, stabilized "Fischer" carbenes and the *nucleophilic*, methylene or alkylidene, "Schrock" carbenes. They are quite different in their reaction chemistry, and will be treated separately. However, it is important to note that these represent bonding extremes, and a continuum of complexes of intermediate reactivity almost certainly exists.

6.2 Electrophilic Heteroatom Stabilized "Fischer" Carbene Complexes[1]

The most intensively developed carbene complexes for use in organic synthesis are the electrophilic Fischer carbene complexes of the Group 6 metals, Cr, Mo, W. These are readily synthesized by the reaction of the air-stable, crystalline metal hexacarbonyl with a range of organolithium

Eq. 6.1

reagents (Eq. 6.1). One of the six equivalent CO groups undergoes attack to produce the stable, anionic lithium acyl "ate" complex, in which the negative charge is stabilized and extensively delocalized into the remaining five, π-accepting, electron withdrawing CO groups. These "ate" complexes are normally isolated as the stable tetramethyl ammonium salt, which can be prepared on a large scale and stored for months without substantial decomposition. Treatment with hard alkylating agents such as methyl triflate or trimethyloxonium salts (methyl Meerwein's reagent) results in alkylation at oxygen, producing the alkoxycarbene complex in excellent yield. These yellow to red crystalline solids are easily purified by crystallization or chromatography on silica gel. As solids they are quite air-stable and easy to handle. In solution, they are slightly air-sensitive, particularly in the presence of light, and reactions are best carried out under an inert atmosphere. The heteroatom is required for stability (the diphenyl carbene complex can be made, but decomposes above –20°C), and this stability results from extensive delocalization of the lone pair on the heteroatom into the strongly electron withdrawing metal pentacarbonyl fragment. This delocalization is evidenced by restricted rotation (14-25 kcal/mol) about the heteroatom-carbene carbon bond, and can be correlated to ^{53}Cr NMR chemical shifts.[2] For delocalization the orbital of the lone pair must be colinear with the p orbital of the carbene carbon; because the

Figure 6.1   The "CO Wall" in Group 6 Carbene Complexes

chromium pentacarbonyl fragment presents the carbene carbon with a "wall" of CO's, α-branching of the substituents on the carbene carbon results in steric congestion which may prevent efficient overlap and compromise the stability of the carbene complex (Figure 6.1). A practical consequence of this is the difficulty experienced in preparing branched carbene complexes for use in the synthesis of complex molecules.

Electrophilic carbene complexes have a very rich chemistry, and undergo reaction at several sites. Fischer carbene complexes are coordinatively saturated, metal(0) d^6 complexes which undergo ligand exchange (CO loss) by a dissociative process (Eq. 6.2). Since loss of CO is a prerequisite for substrate coordination and subsequent reaction, this exchange process is central to most synthetically useful reactions. It can be driven thermally ($T_{1/2}$ for CO exchange = 5 minutes at 140°C) or photochemically, and most organic reactions of carbenes involve one of these modes of activation.

Eq. 6.2

$(M^0, d^6, 18e^-, sat)$

Because CO groups are strongly electron withdrawing, the metal-carbene carbon bond is polarized such that the carbene carbon is electrophilic, and generally subject to attack by a wide range of sterically unencumbered nucleophiles (Eq. 6.3). In most cases the resulting tetrahedral intermediate is unstable, and the alkoxy group is ejected, producing a new carbene complex. This is one of the best ways to prepare Fischer carbene complexes containing heteroatoms other than oxygen, and the process is quite analogous to "transesterification" of organic esters. In fact, in many of their reactions the analogy between alkoxycarbene complexes and organic esters is remarkable.

Eq. 6.3

$$Nuc = R'O^-, NH_3, RNH_2$$
$$\underline{Small}\ R_2NH, RSH$$

Protons α- to the carbene carbon are quite acidic ($pK_a \approx 12$) and can be removed by a variety of bases to give an "enolate" anion stabilized by delocalization into the metal carbonyl fragment (Eq. 6.4).[3] These anions are only weakly nucleophilic, but in the presence of Lewis acid catalysts they react with epoxides (Eq. 6.5)[4] and aldehydes (Eq. 6.6)[5] to produce homologated carbene complexes. These reactions are important, since they allow elaboration of carbene fragments for ultimate incorporation into organic substrates. Alkylation of these anions with active halides often results in messy polyalkylation. Use of triflates mitigates this problem, as does replacing one of the acceptor CO ligands with a donor phosphine ligand.[6] This makes the chromium fragment a poorer acceptor, decreases the acidity of the α-protons and increases the reactivity of the anion. Similarly, aminocarbene complexes produce more reactive α-anions than do alkoxycarbenes (compare the α-proton acidity of amides vs. esters, N being a stronger donor than O), and are more easily α-alkylated (see below). With epoxides, the γ-hydroxy compound is isolated, since alkoxy groups will not displace amino groups (Eq. 6.7).[7]

Eq. 6.4

Eq 6.5

Eq. 6.6

Eq. 6.7

R = H, Me, Et, CH₂Cl

30-70%

Strong bases which irreversibly deprotonate carbene complexes are often required. Weaker bases such as pyridine catalyze the decomposition of alkoxycarbene complexes into enol ethers, by reversible α-deprotonation, followed by reprotonation on the metal and reductive-elimination (Eq. 6.8). This reaction could be used (although it has not) to synthesize complex enol ethers. However, it has been used to synthesize optically active ene carbamates (Eq. 6.9).[8] In this case, the putative N-acyl carbene intermediate is destabilized both by N-acylation and by steric hindrance to planarity for efficient π overlap of the lone pair on the hetero atom with the carbene carbon. As a consequence α-deprotonation becomes favored, and the N-acyl carbene complex has a fleeting (but detectable) lifetime.

Eq. 6.8

Eq. 6.9

α,β-Unsaturated carbene complexes, prepared by aldol condensation (Eq. 6.6) or from vinyllithium reagents, undergo many reactions common to unsaturated esters but are considerably more reactive. For instance, Diels-Alder reactions are 2×10^4 faster with carbene complexes than with esters, permitting very facile elaboration of carbene complexes (Eq. 6.10).[9] This, again, is a consequence of the electron withdrawing ability of the $Cr(CO)_5$ fragment. Michael addition reactions to conjugated carbene complexes are also efficient. In fact, in the exchange of the alkoxy group by amines in α,β-unsaturated carbene complexes, Michael addition of the amine is the major competing pathway. Michael addition provides a very useful route for elaborating carbene complexes, for ultimate use in synthesis. Stabilized enolates add readily, and the resulting carbene anion is inert to further attack by nucleophiles, permitting further functionalization of the newly introduced carbonyl group (Eq. 6.11).[10] Similarly, acetylenic carbene complexes are much more reactive to the Pauson-Khand process (Chapter 8), permitting the construction of elaborate carbene complexes under very mild conditions (Eq. 6.12).[11]

Eq. 6.10

Eq. 6.11

Eq. 6.12

80%

Once elaborated, the carbene fragment can be converted to organic products in a number of ways. Oxidation with Ce(IV) or dimethyldioxirane (Eq. 6.13)[12] generates the carboxylic acid derivative, as does oxidation of the tetramethylammonium salt in the presence of a nucleophile.[13] However, the metal-carbon double bond can be, and has been, put to better use.

Eq. 6.13

R = Me, Ph, Bu, Ph —≡

X = OEt, OMe, NHPr

One of the first organic reactions of Group 6 carbene complexes studied was the cyclopropanation of electrophilic olefins (Eq. 6.14).[14] The reaction is carried out by simply heating mixtures of the carbene complex and olefin, although often at elevated temperatures. Moderate to good yields of the cyclopropane are obtained. With 1,2-disubstituted alkenes, the stereochemistry of the alkene is maintained in the product, but mixtures of stereoisomers at the (former) carbene carbon are obtained. Dienes, if they are not too highly substituted, are monocyclopropanated (Eq. 6.15)[15] and (Eq. 6.16).[16] In many cases, in addition to the cyclopropane product, varying amounts of a by-product corresponding to vinyl C–H insertion are obtained (Eq. 6.17).[17]

Eq. 6.14

$Z = CO_2Me, CONMe_2, CN, P(O)(OMe)_2, SO_2Ph$

Eq. 6.15

Eq. 6.16

Eq. 6.17

Although cyclopropanes are the expected product of the reaction of free organic carbenes and olefins, there is ample evidence that free organic carbenes are not involved in the cyclopropanation reactions of transition metal carbene complexes. Rather the route in Figure 6.2 is followed. Cyclopropanation is suppressed by CO pressure, indicating that loss of CO to generate a vacant coordination site is required. (a) Coordination of the alkene, (b) followed by formal 2+2 cycloaddition to the metal-carbon double bond (c) would generate a metallocyclobutane. Depending on the regioselectivity of this cycloaddition step, the alkene substituent may end up either α or β to the metal. Reductive elimination (d) gives the cyclopropane. The observed "C–H" insertion by-product could be formed by β-hydride elimination (e) then reductive elimination (f) from the metallacyclobutane.

Evidence for metallacyclobutane intermediates comes from attempted cyclopropanation of electron-rich alkenes. Under normal reaction conditions, cyclopropanation does not result; rather a new alkene resulting from metathesis (alkene group interchange) is obtained (Eq. 6.18).

Figure 6.2   Mechanism of Cyclopropanation

Eq. 6.18

Olefin metathesis is a well known process for carbene complexes, and has found many industrial uses. It is thought to occur by the cycloaddition/cycloreversion process shown. Because metathesis of electron rich olefins produces heteroatom stabilized carbene complexes, whereas electron poor olefins would produce unstabilized carbene complexes, metathesis is the favored pathway with electron rich substrates. Electron rich alkenes can be cyclopropanated by unstable acyloxycarbene complexes, generated *in situ* by O-acylation of tetramethyl ammonium (but not lithium, because of tight ion pairing) "ate" complexes (Eq. 6.19),[18] or under high (>100 atm) pressures of CO, which seems to suppress metathesis.

In contrast to electron poor and electron rich alkenes, simple aliphatic alkenes do not undergo intermolecular cyclopropanation by chromium alkoxycarbenes under any conditions. However, in general, intramolecular reactions are considerably more favorable, and aryl-*alken*oxycarbene complexes undergo extremely facile cyclopropanation of the unactivated alkene (see below).

Eq. 6.19

Carbene complexes with structurally complex alkoxy groups are not directly available by the standard synthesis (Eq. 6.1) because the corresponding alkyl triflates or oxonium salts are difficult to synthesize. Further, they cannot be made by alkoxy-alkoxy' interchange, since these reactions (in contrast to alkoxy-amine exchange) are slow, inefficient, and beset with side reactions. However, acyloxycarbene complexes, generated *in situ* by reaction of acid halides with ammonium acylate complexes, react cleanly, even with structurally complex alcohols, to produce alkoxy-carbene complexes in excellent yield (Eq. 6.20).[19] When homoallylic alcohols were exchanged into (aryl)(acyloxy) carbenes, the carbene complex

Eq. 6.20

(won't O-acylate)

"mixed anhydride"

could not be isolated. Instead, spontaneous, intramolecular cyclopropanation occurred, even under modest pressures of carbon monoxide (Eq. 6.21).[20] This is a general phenomenon, and a variety of cyclopropanated tetrahydrofurans, pyrans (Eq. 6.22) and pyrrolidines (Eq. 6.23) are available by this route. This increase in reactivity is clearly entropic; when the alkene is more than three carbons removed from the oxygen, cyclopropanation does not occur under conditions sufficiently severe to decompose the complex.

Eq. 6.21

Eq. 6.22

(can be isolated)

92%

Eq. 6.23

61%

This intramolecular cyclopropanation process offers additional evidence for the intermediacy of metallacyclobutanes, and the potential competition by metathesis, in these cyclopropanation reactions. In contrast to most cyclopropanations, heating either the *cis* or the *trans* 3-hexenyloxycarbene complex produced a mixture of stereoisomers of bicyclic cyclopropanes, along with small amounts of a dihydrofuran (Eq. 6.24). This loss of stereochemistry is best accommodated by the intervention of a metathesis/recycloaddition step, which can both scramble stereochemistry and produce the dihydrofuran by-product.

Eq. 6.24

E	52%	20%	3%
Z	22%	32%	

via

Related metathesis/recycloaddition sequences have been used to expeditiously synthesize bicyclic compounds. Alkynes cycloadd to carbene complexes more readily than do alkenes. Treatment of enynes with chromium carbene complexes produced bicyclic compounds in excellent yield, through a process of alkyne cycloaddition/metathesis/alkene cycloaddition/reductive elimination (Eq. 6.25).[21] Many, many possible variations of this process exist, and several have been tried (Eq. 6.26[22] and Eq. 6.27[23]). All lead to a remarkable increase in complexity in a single step.

Eq. 6.25

≈90%

(a) X = C(CO$_2$Et)$_2$, Z = N⟩ (b) X = NTs, Z = OEt

Eq. 6.26

60%

Eq. 6.27

Another thermal reaction of carbene complexes of great use in synthesis is the Dötz reaction, the reaction of unsaturated alkoxycarbene complexes with alkynes, to produce hydroquinone derivatives. The general transformation is shown in Eq. 6.28.[24] Although the overall sequence is complex, the mechanism is reasonably well-understood, and involves many of the standard processes of organometallic chemistry (Eq. 6.29).

Eq. 6.28

Eq. 6.29

Cr-vinyl ketene complex

carbene

alkyne

Again, the reaction begins with a thermal loss of one CO, to generate a vacant coordination site. Coordination of the alkyne followed by cycloaddition generates the metallacyclobutene, which inserts CO to give a metallacyclopentenone. Fragmentation of this metallacycle generates a metal-bound vinyl ketene, for which there is excellent experimental evidence (X-ray in one case, isolation of the vinylketene with sterically hindered alkynes[25]). Cyclization of this vinyl ketene followed by enolization produces the hydroquinone derivative, as its Cr(CO)$_3$ complex (Chapter 10). The free ligand is obtained by oxidative removal of chromium by exposure to air and light.

Because of the stability and ease of handling of Group 6 carbene complexes, and the many ways they can be elaborated, highly functionalized, structurally complex carbene complexes are relatively easily prepared, making their use in organic synthesis appealing. The Dötz reaction has been extensively utilized to make quinone derivatives. Selected examples are shown in Eq. 6.30,[26] Eq. 6.31,[27] Eq. 6.32,[28] and Eq. 6.33.[29] As is the case with cyclopropanation, a cascade of different reactions can be arranged so as to make many bonds in a single reaction sequence. Such a case is seen in Eq. 6.34,[30] in which a Diels-Alder cycloaddition is followed by an alkyne metathesis, and then a Dötz reaction, all in a one-pot reaction.

Eq. 6.30

Eq. 6.31

Eq. 6.32

Eq. 6.33

X = O, NMe, CH=CH

Eq. 6.34

As can be seen, the Dötz reaction has many steps, each of which can be diverted along other pathways, leading to other products. The main problem in the development of this chemistry was to develop conditions which led to a single major product, since the reaction is quite sensitive to both the reaction conditions and structural features of the reactants.

For example, when cyclopropylcarbenes are subjected to the typical Dötz reaction, totally unexpected products result (Eq. 6.35).[31] Although the mechanism is not well-understood, it is likely that things proceed as usual to the metallacyclopentenone stage (b), at which point, the *"sp2"* like cyclopropyl group inserts as if it were an olefin (cyclopropyl groups are readily cleaved by a variety of transition metals) to give a metallacyclooctadienone (c). This then reinserts to give a bicyclic system (d)(shown as occurring in two steps, for clarity), which fragments to give the cyclopentadienone and an alkene (e). The dienone is reduced by the low valent chromium fragment in the presence of water. Intramolecular versions also work (Eq. 6.36).[32] When the metal is changed from chromium to the larger tungsten, seven-membered rings form, apparently from direct reductive elimination from the metallacyclooctadienone (Eq. 6.37).[33] (Note that isomerization of the double bonds via 1,5-hydride shifts to give the most stable product occurs under the relatively severe reaction conditions.)

Eq. 6.35

40-80%

Eq. 6.36

30-60%

Eq. 6.37

45-65%

65%

The Dötz reaction is also quite sensitive to the nature of the heteroatom on the carbene complex. Although alkoxycarbene complexes give hydroquinone derivatives, aminoquinones fail to insert CO, and give indane derivatives instead (Eq. 6.38).[34] When carried out intramolecularly, the reaction is remarkably diastereoselective (Eq. 6.39).[35]

Eq. 6.38

Eq. 6.39

~50%

Iron carbene complexes analogous to the Group 6 Fischer carbenes are easily prepared, either by the Fischer route (Eq. 6.1) or from Collman's reagent ($Na_2Fe(CO)_4$) (Eq. 6.40). In the thermal reaction with alkenes, no cyclopropenes are formed; rather, C–H insertion products predominate (Eq. 6.41).[36] With alkynes, iron alkoxycarbenes react by a completely different route from Group 6 analogs, inserting *two* CO's and giving a mixture of equilibrating pyrones (Eq. 6.42)![37]

Eq. 6.40

Eq. 6.41

40-60%

Eq. 6.42

(X-ray)

In addition to thermal reactions, chromium heteroatom carbene complexes have a very rich photochemistry which results in a number of synthetically useful processes. All of these complexes are colored and have absorptions in the 350-450 nm range. This visible absorption has been assigned as an allowed metal-to-ligand charge transfer (MLCT) band.[38] From the molecular orbital diagram of the complex (Figure 6.3), the HOMO is metal-d-orbital centered, and the LUMO is carbene carbon p-orbital centered. Thus, photolysis into the MLCT results in a *formal*, reversible, one electron oxidation of the metal, since, in the absorption pro-

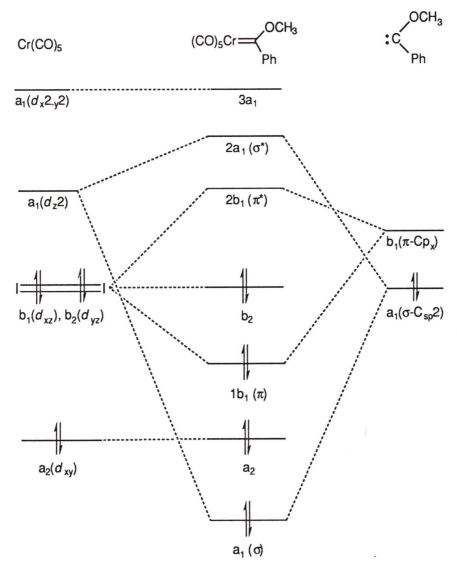

Figure 6.3 Molecular Orbital Diagram for Chromium Carbene Complexes

cess, an electron is removed from the *d*-centered HOMO to the *p* centered LUMO.[39] One of the best ways to drive insertion of carbon monoxide into metal-carbon bonds is to oxidize the metal (Chapters 2 and 4) and, indeed, photolysis of chromium carbene complexes with visible light appears to drive a reversible insertion of CO into the metal carbon double bond, producing a metallacyclopropanone, best represented as a metal-bound ketene (Eq. 6.43).[40] Since excited states of transition metal complexes are usually too short-lived to undergo intermolecular reactions, this is almost surely a ground-state complex which, if not trapped, deinserts CO, regenerating the carbene complex. Thus, many carbene complexes can be recovered unchanged after extended periods of photolysis in the absence of reagents which react with ketenes, but are completely consumed in a few hours when photolyzed in the presence of such reagents.

Eq. 6.43

From a synthetic point of view, this is very useful, since it permits the generation of unusual, electron rich alkoxy, amino, and thioketenes not easily available by conventional routes, under very mild conditions (visible light, any solvent, Pyrex vessels, no additional reagents). The ketene is metal bound, and generated in low concentrations, thus normal ketene side reactions such as dimerization, or incorporation of more than one ketene unit into the product, are not observed. However, the species thus generated undergo most of the normal reactions of ketenes, and herein lies their utility.

The reaction of ketenes with imines to produce β-lactams (the Staudinger reaction) has been extensively developed for the synthesis of this important class of antibiotics. Most biologically active β-lactams are optically active, and have an amino group α- to the lactam carbonyl. To prepare these from chromium carbene complex chemistry, an optically active complex having a hydrogen on the carbene carbon is necessary (Eq. 6.44). These are not available from conventional Fischer synthesis, since anionic metal formyl complexes are strong reducing agents and transfer hydrides to electrophiles. The requisite optically active aminocarbene complex can be made efficiently by an alternative route involving the reaction of $K_2Cr(CO)_5$ with the appropriate amide, followed by addition of TMSCl to promote the displacement of the oxygen (Eq. 6.45).[41]

Eq. 6.44

Eq. 6.45

This optically active aminocarbene complex undergoes efficient reaction with a very wide range of imines to give optically active β-lactams in excellent chemical yield, and with very high diastereoselectivity (Eq. 6.46).[42] The absolute stereochemistry of the α-position is determined by the chiral auxiliary on nitrogen, which transfers its absolute configuration to this position (R → R, S → S), while the relative (cis-trans) stereochemistry between the two new chiral centers is determined by the imine.[43] When the reaction is carried out under a modest (60 psi) pressure of CO, chromium hexacarbonyl can be recovered and reused. The chiral auxiliary is easily removed by hydrolysis of the acetonide, followed by either reductive (H_2, Pd/C) or oxidative (IO_4^-) cleavage of the amino alcohol.

This chemistry has been used in a formal synthesis of (+) 1-carba-cephalothin (Eq. 6.47).[44]

Eq. 6.46

Eq. 6.47

Ketenes also undergo facile 2+2 cycloaddition reactions with olefins, producing cyclobutanones, with a high degree of stereoselectivity, favoring the more hindered cyclobutanone ("masochistic stereoinduction"[45]). Photolysis of chromium alkoxycarbene complexes in the presence of a wide variety of alkenes produces cyclobutanones in high yield, with roughly the same stereoselectivity observed using conventionally generated ketenes (Eq. 6.48).[46] As expected, the reaction is restricted to relatively electron rich olefins, and electron deficient olefins fail altogether. It is also restricted to alkoxycarbene complexes or aminocarbene complexes having aryl groups on nitrogen to decrease the nucleophilicity of the ketene.[47] Intramolecular versions also work well, giving access to a variety of bicyclic systems (Eq. 6.49).[46] Even unconventional cycloadditions proceed in remarkably good yield (Eq. 6.50).[48] With optically active alkenes, high diastereoselectivity is observed (Eq. 6.51).[49] The resulting optically active cyclobutanones undergo Baeyer-Villager oxidation/elimination to produce butenolides, useful synthetic intermediates.

Eq. 6.48

Eq. 6.49

Eq. 6.50

95%

Eq. 6.51

Ketenes react with alcohols or amines to give carboxylic acid derivatives. Similarly, photolysis of chromium carbene complexes in the presence of alcohols produces esters in excellent yield. Use of aminocarbene complexes, prepared from amides (Eq. 6.45) provides a very direct synthesis of α-amino acid derivatives from amides, a transformation not easily achieved using conventional methodology (Eq. 6.52).[50] Lactams, including β-lactams, undergo this transformation as well, producing cyclic amino acids.

Eq. 6.52

Optically active α-amino acids are an exceptionally important class of compounds, not only because of their central biological role as the fundamental units of peptides and proteins, but also because of their role in the development of new pharmaceutical derivatives. In the above racemic synthesis of α-amino esters, the stereogenic center is formed by protonation of a ketene derived enolate. To develop an asymmetric synthesis based on this chemistry, efficient asymmetric protonation of this enolate must be achieved. Photolysis of the same optically active aminocarbene complex used to synthesize β-lactams, in methanol d₁ produced "optically active" glycine in excellent chemical yield, with high deuterium incorporation, and in 86% ee (Eq. 6.53).[51] Since this is the sterically least demanding asymmetric protonation, the process is even more diastereoselective for making higher homologs (alanine, etc.) (Eq. 6.54).[50] By taking advantage of the α-acidity of carbene complexes (Eq. 6.4-Eq. 6.7) a variety of homologated alanines are available by this route.

Eq. 6.53

Eq. 6.54

By replacing the alcohol with an optically active α-amino acid ester, dipeptides can be synthesized in a process that forms the peptide bond and the new stereogenic center on the carbene-complex-derived amino acid fragment in a single step (Eq. 6.55).[52] This permits the direct introduction of natural (or unnatural) amino acid fragments into peptides utilizing visible light as the coupling agent. In this case, "double diastereoselection" is observed with the (R,S,S) dipeptide being the matched and the (S,R,S) dipeptide being the mismatched pair.

Eq. 6.55

Finally, combining the cyclization of vinyl ketenes with adjacent unsaturation (the Dötz reaction Eq. 6.29), with the photochemical generation of ketenes from carbene complexes, provides a very clever synthesis of ortho quinone derivatives (Eq. 6.56).[53]

Eq. 6.56

6.3 Electrophilic, Nonstabilized Carbene Complexes[54]

Group 6 carbene complexes lacking a heteroatom on the carbene carbon are unstable and difficult to prepare. As a consequence, they have not been extensively utilized in synthesis (although they do cyclopropanate olefins[55]). In contrast, a number of stable, easily prepared iron carbonyl complexes are precursors to unstabilized electrophilic carbene complexes.

When these precursor complexes are treated with appropriate reagents in the presence of alkenes, cyclopropanation results. Among the most convenient of these carbene precursors are the methylthiomethyl complexes, easily prepared on large scale by the reaction of $CpFe(CO)_2Na$ with CH_3SCH_2Cl (Eq. 6.57).[56] S-Methylation followed by warming in the presence of an excess of electron rich alkene produces the unstable cationic carbene complex which then cyclopropanates the alkene. The corresponding methoxymethyl complex behaves similarly. Another, less convenient approach is shown in Eq. 6.58. Perhaps the most versatile approach relies upon the reaction of the strongly nucleophilic $CpFe(CO)_2Na$ with aldehydes[57] or ketals[58] to produce the requisite α-oxoiron complexes. These undergo α-elimination to produce cationic carbenes when treated with alkylating or silylating agents (Eq. 6.59).

Eq. 6.57

Eq. 6.58

Eq. 6.59

This cyclopropanation is somewhat limited. Only relatively un-functionalized, electron rich alkenes undergo the reaction, and relatively few carbene groups — "CH_2", "$CHCH_3$", "$C(CH_3)_2$", "CHPh", "CHC_3H_5" — can be transferred. With substituted carbene complexes, mixtures of *cis* and *trans* disubstituted cyclopropanes are obtained.

By replacing one of the carbon monoxide groups with a phosphine, the iron carbene complex precursors become chiral at iron, and can be re-solved. Additional asymmetry can be introduced by using an optically active phosphine, giving complexes chiral both at iron and phosphorus. These optically active carbene complexes cyclopropanate alkenes with quite good enantioselectivity, but only modest *cis/trans* selectivity (Eq. 6.60). The mechanism of this asymmetric cyclopropanation has been extensively studied.[59] Because chiral metal complexes induce asymmetry in the products, "free" carbenes cannot be involved. Unexpectedly perhaps, metallacyclobutanes also are not involved. From careful stereochemical and labeling studies, the mechanism shown in Eq. 6.61 has been estab-lished. It involves attack of the alkene on the electrophilic carbene com-plex to generate an electrophilic center at the γ-carbon. If this carbon bears an electron donating group, this intermediate is sufficiently long-lived to permit rotation about the β-γ-bond, leading to loss of the original olefin stereochemistry. This γ-cationic center is then attacked by the Fe-$C_α$ bond from the back side resulting in inversion at this center.

Eq. 6.60

Eq. 6.61

6.4 Metal-Catalyzed Decomposition of Diazo Compounds Proceeding Through Unstabilized Electrophilic Carbene Intermediates

Transition metals catalyze a number of important reactions of organic diazo compounds, among which cyclopropanation of olefins is one of the most synthetically useful (Eq. 6.62).[60] A wide variety of transition-metal complexes are efficient catalysts, among which copper(I) triflate, palladium(II) salts, rhodium(II) acetate, and $Rh_6(CO)_{16}$ are most commonly used. The reaction is most effective when relatively stable α-diazocarbonyl compounds are used, since the major competing reaction is decomposition of the diazo compound itself. Olefins ranging from electron-rich enol ethers to electron-poor α,β-unsaturated esters undergo catalytic cyclopropanation with diazo compounds, although alkenes with electron-withdrawing substituents are likely to react by a mechanism different from that of other olefins, and β-substitution drastically suppresses the reaction. The order of reactivity for olefins is electron rich > "neutral"

>> electron poor and *cis* > *trans*. Olefin geometry is maintained, but there is little stereoselectivity for the diazo-derived group. Dienes and alkynes are also cyclopropanated. Detailed mechanistic studies have implicated the intermediacy of metal carbene complexes in this cyclopropanation (Eq. 6.62).[61]

Eq. 6.62

$ML_n = Rh_2(OAc)_4, CuCl \cdot P(OiPr)_3, Rh_6(CO)_{16}, PdCl_2 \cdot 2PhCN$

$X = BrCH_2, ClCH_2, PhO, nBu, OAc, OEt, O-nBu, iPr, tBu, CH_2=CPh$

$CH_2=CMe, CH_2=C-tBu, CH=CHOMe, CH=CHCl, CH=CHPh, CH=CHMe$

From a synthetic point of view, rhodium(II) acetate is by far the most-used catalyst, because it is commercially available, easy to handle, and reacts quickly and efficiently. Many functional groups are tolerated making elaborate cyclopropanes available (for example Eq. 6.63).[62]

Eq. 6.63

W = Me, H, OSiR₃
X = Me, H
Y = H, OAc, OMe, OSiR₃
Z = H, OAc

Intramolecular versions are quite efficient, producing polycyclic ring systems (Eq. 6.64[63] and Eq. 6.65[64]). By using rhodium(II) complexes of optically active carboxylates or carboxamides, highly asymmetric induction can be achieved (Eq. 6.66).[65,66]

Eq. 6.64

Eq. 6.65

83% yield   69:31 *cis/syn* OR:*cis/anti* OR

Eq. 6.66

72-94% ee
30-88% yield

R^1 = H, Me, Ph, Pr   R^2 = H, Me, Ph, Et, Bn, ⬡ , *i*Bu, Bu$_3$Sn

When substrates have additional unsaturation appropriately situated, metathesis "cascade" cyclizations ensue (Eq. 6.67),[67] much like the corresponding reactions of chromium carbene complexes (Eqs. 6.25-6.27). This provides further evidence for the intermediacy of metal carbene complexes in these cyclopropanations, since such metathesis derived products are unlikely for other types of carbenoid species.

Eq. 6.67

95%!

As stated above, a very large number of transition metals catalyze the cyclopropanation of alkenes by diazocompounds. A particularly impressive example using palladium(II) is shown in Eq. 6.68.[68] Copper triflate or perchlorate also catalyzes cyclopropanation. In the presence of optically active bidentate heterocycles very good asymmetric induction is achieved (Eq. 6.69).[69]

Eq. 6.68

87%

Eq. 6.69

L* =

> 95% *trans*
94% ee

Cyclopropanation of alkenes by metal-catalyzed decomposition of diazoalkanes has many parallels to related reactions of stable carbene complexes, and the two systems are clearly mechanistically related. However, two reactions not common for metal carbene complexes — X–H insertion and ylide formation — are observed with metal catalyzed decomposition of diazoalkanes, and both are synthetically useful.

Rhodium(II) catalyzes the insertion of diazoalkanes into C–H, O–H, N–H, and S–H bonds in a process that is very unlikely to involve free carbenes (see below) but for which detailed mechanisms involving metal carbene complexes, have not yet been elucidated. The C–H insertion process can be quite synthetically useful (Eq. 6.70[70] and Eq. 6.71[71]). C–H insertion is a typical reaction for free organic carbenes, but the metal must be involved here, since the site of insertion can be altered by changing the ligands on the metal. For example, the system in Eq. 6.72 gives *only* fused bicyclic products when rhodium(II) tritylcarboxylate is the catalyst, and mainly spirofused products when rhodium(II) acetate is used.[72] Even more compelling evidence for the role of the metal in these C–H insertions comes from the high degree of asymmetric induction observed when optically active metal catalysts are used (Eq. 6.73)[73] and (Eq. 6.74)[74] implying an intimate association of the metal with the carbene fragment during bond formation. Rhodium(II) salts also catalyze N–H (Eq. 6.75)[75] and O–H (Eq. 6.76)[76] insertions which are synthetically useful.

Eq. 6.70

70-80%

Eq. 6.71

100%

Eq. 6.72

≈70%

Eq. 6.73

40-80%
45-91% ee

Eq. 6.74

75-80% yield
62-82% ee

Eq. 6.75

Eq. 6.76

71%

Finally, transition metals catalyze the reaction between diazo compounds and heteroatom lone pairs to produce ylides which have a very rich chemistry in their own right.[77] The general process is shown in Eq. 6.77, and *formally* consists of attack of the lone pair on the electron deficient carbene carbon to give the ylide, which then undergoes further reaction. There are now literally hundreds of examples of this chemistry, and very complex structures can be rapidly assembled (Eq. 6.78, Eq. 6.79, and Eq. 6.80[78]).

Eq. 6.77

Eq. 6.78

Eq. 6.79

(carbonyl ylide)

1,3
dipolar
cycloaddition

Eq. 6.80

n = 1-3   83-88%

6.5 Nucleophilic "Schrock" Carbene Complexes

Nucleophilic carbene complexes represent the opposite extreme of reactivity and bonding from electrophilic carbene complexes. They are usually formed by metals at the far left of the transition series, in high oxidation states (usually d^0) with strong donor ligands, such as alkyl or cyclopentadienyl, and no acceptor ligands. The carbene ligand is usually simply the $=CH_2$ group. These are so different from electrophilic carbene complexes that the carbene ligand in these complexes is formally considered a four electron, dinegative ligand, $[H_2C::]^{2-}$. These complexes often have Wittig-reagent-like reactivity and are often represented as $M^+-CH_2^- \leftrightarrow M=CH_2$, much like ylides. However, nucleophilic carbene complexes also form metallacycles, and both of these aspects of reactivity are useful in synthesis.

The most extensively studied nucleophilic carbene complex is "Tebbe's" reagent, prepared by the reaction of trimethyl aluminum with titanocene dichloride (Eq. 6.81).[79] In the presence of pyridine this complex is synthetically equivalent to "Cp_2TiCH_2", and it is very efficient in converting carbonyl groups to methylenes (Eq. 6.82).[80] The reaction is thought to proceed via an oxometallacycle which can fragment to give the alkene and a very stable titanium(IV) oxo species (Eq. 6.83). The reaction not only works well with aldehydes and ketones, but it also converts esters to enol ethers and amides to enamines! With acid halides and anhydrides, titanium enolates are produced instead. Tebbe's reagent does not enolize ketones and is particularly efficient for methylenating sterically hindered, enolizable ketones which are not methylenated efficiently by Wittig reagents.[81] The reagent tolerates a very high degree of functionality (Eq. 6.84[82], Eq. 6.85[83], and Eq. 6.86[84]), and does not epimerize optically active systems (Eq. 6.87[80] and Eq. 6.88[85]).

Eq. 6.81

Eq. 6.82

$$X = R', H, OR', NR'_2$$

Eq. 6.83

Eq. 6.84

92%

Eq. 6.85

90%

Eq. 6.86

69%

Eq. 6.87

Eq. 6.88

94%

An even more easily prepared methylenating reagent is dimethyl titanocene, Cp_2TiMe_2, generated by the simple treatment of titanocene dichloride with methyllithium. Gentle heating of this complex (~60°) in the presence of ketones, aldehydes, or esters results in methylation with yields and selectivities similar to those observed with Tebbe's reagent.[86] In some instances, it has advantages over Tebbe's reagent. In the system in Eq. 6.89, Tebbe's reagent fragmented the substrate while Cp_2TiMe_2 cleanly methylenated it.[87] In addition, benzylidene (PhCH=)[88] and trimethylsilylmethylene (TMSCH=)[89] can be transferred, while Tebbe's reagent is restricted to simple methylene groups.

Eq. 6.89

high yield

Another very simple to prepare methylenating reagent is prepared from zirconocene dichloride (Eq. 6.90).[90] This reagent methylenates aldehydes and conjugated ketones, but has not been extensively studied.

Eq. 6.90

In addition to methylenating carbonyl compounds, Tebbe's reagent reacts with olefins to form titanacycles. These can be cleaved by a variety of reagents (Eq. 6.91), or can be used to synthesize allenes by a cycloaddition/metathesis sequence (Eq. 6.92).[91]

Eq. 6.91

Eq. 6.92

References

1. Seyferth, D., ed. *Transition Metal Carbene Complexes*, Verlag Chemie, 1983.
2. Hafner, A.; Hegedus, L.S.; deWeck, G.; Hawkins, B.; Dötz, K.H. *J. Am. Chem. Soc.* **1988**, *110*, 8413.
3. Casey, C.P., "Metal-Carbene Complexes in Organic Synthesis", in *Transition Metal Organometallics in Organic Synthesis*, vol. 1, Alper, H. ed. Academic Press, New York, 1976, pp. 190-23.
4. Lattuada, L.; Licandro, E.; Maiorana, S.; Molinari, H.; Papagni, A. *Organometallics* **1991**, *10*, 807.
5. Aumann, R.; Heinen, H. *Chem. Ber.* **1987**, *120*, 537.
6. Xu, Y-C.; Wulff, W.D. *J. Org. Chem.* **1987**, *52*, 3263.
7. Lattuada, L.; Licandro, E.; Maiorana, S.; Papagni, A.; Zenotta-Gerosa, A. *Syn. Lett.* **1992**, 315.
8. Montgomery, J.; Wieber, G.M.; Hegedus, L.S. *J. Am. Chem. Soc.* **1990**, *112*, 6255.
9. Wang, S.L.B.; Wulff, W.D. *J. Am. Chem. Soc.* **1990**, *112*, 4550. Wulff, W.D.; Bauta, W.E.; Kaesler, R.W.; Lankford, P.J.; Miller, R.A.; Murray, C.K.; Yang, D.C. *J. Am. Chem. Soc.* **1990**, *112*, 3642.
10. Aoki, S.; Fujimura, T.; Nakamura, E. *J. Am. Chem. Soc.* **1992**, *114*, 2985.
11. Dötz, K.H.; Christoffers, J. *J. Organomet. Chem.* **1992**, *426*, C58; for preparation of stable alkynyl $Co_2(CO)_6$ complexes see: Wienand, A.; Reissig, H-U.; Fischer, H.; Pflumm, D.; Troll, C. *J. Organomet. Chem.* **1992**, *427*, C9.
12. Lluch, A.M.; Jordi, L.; Sanchez-Baeza, F.; Ricart, S.; Camps, F.; Messeguer, A.; Moreto, J.M. *Tetrahedron Lett.* **1992**, *33*, 3021.
13. Soderberg, B.C.; Bowden, B.A. *Organometallics* **1992**, *11*, 2220.
14. For reviews see: (a) Dötz, K.H. *Angew. Chem. Int. Ed. Engl.* **1984**, *23*, 587. (b) Reissig, H-U. *Organomet. in Synthesis* **1989**, *2*, 311. For recent studies see: Reissig, H-U. *Organometallics* **1990**, *9*, 3133.
15. Harvey, D.F.; Lund, K.P. *J. Am. Chem. Soc.* **1991**, *113*, 8916.
16. Fischer, H.; Hofmann, J. *Chem. Ber.* **1991**, *124*, 981.
17. Wienand, A.; Reissig, H.U. *Angew. Chem. Int. Ed. Engl.* **1990**, *29*, 1129.
18. Murray, C.K.; Yang, D.C.; Wulff, W.D. *J. Am. Chem. Soc.* **1990**, *112*, 5660.
19. Semmelhack, M.F.; Bozell, J.J. *Tetrahedron Lett.* **1982**, *23*, 2931.
20. Soderberg, B.C.; Hegedus, L.S. *Organometallics* **1990**, *9*, 3113.
21. (a) Hoye, T.R.; Suriano, J.A. *Organometallics* **1992**, *11*, 2044; **1989**, *8*, 2670. (b) Mori, M.; Wantanuki, S. *J. Chem. Soc., Chem. Comm.* **1992**, 1082. (c) See also: Katz, T.J. *Tetrahedron Lett.* **1991**, *32*, 5895.
22. Harvey, D.F.; Brown, M.F. *J. Am. Chem. Soc.* **1990**, *112*, 7806. *Tetrahedron Lett.* **1991**, *32*, 2871. *Tetrahedron Lett.* **1991**, 5223, 6311.
23. Alvarez, C.; Parlier, A.; Rudler, H.; Yefsah, R.; Daran, J.C.; Knobler, C. *Organometallics* **1989**, *8*, 2253.
24. Dötz, K.H. *Angew. Chem. Int. Ed. Engl.* **1984**, *23*, 587.
25. Dötz, K.H.; Muhlemeier, J. *Angew. Chem. Int. Ed. Engl.* **1982**, *21*, 929.
26. Boger, D.; Jacobson, I.C. *J. Org. Chem.* **1991**, *56*, 2115.
27. Parker, K.A.; Coburn, C.A. *J. Org. Chem.* **1991**, *56*, 1666.
28. Bauta, W.E.; Wulff, W.D.; Pavkovic, S.F.; Zaluzec, E.J. *J. Org. Chem.* **1989**, *54*, 3249.
29. Yamashita, A.; Toy, A.; Scahill, J.A. *J. Org. Chem.* **1989**, *54*, 3625.
30. Bao, J.; Dragisich, V.; Wenglowsky, S.; Wulff, W.D. *J. Am. Chem. Soc.* **1991**, *113*, 9873.
31. Herndon, J.W.; Tumer, S.U. *J. Am. Chem. Soc.* **1988**, *110*, 3334. *Tetrahedron Lett.* **1989**, *30*, 295.
32. Herndon, J.W.; Matasi, J.J. *J. Org. Chem.* **1990**, *55*, 786.

33. Herndon, J.W.; Chatterjee, G.; Patel, P.P.; Matasi, J.J.; Tumer, S.U.; Harp, J.J.; Reid, M.D. *J. Am. Chem. Soc.* **1991**, *113*, 7808.

34. (a) Yamashita, A. *Tetrahedron Lett.* **1986**, *27*, 5915. For reviews on aminocarbene complex chemistry se: (b) Grotjahn, D.B.; Dötz, K.H. *Synlett* **1991**, 381. (c) Schwindt, M.P.; Miller, J.R.; Hegedus, L.S. *J. Organomet. Chem.* **1991**, *413*, 143.

35. Dötz, K.H.; Schäfer, T.O.; Harms, K. *Synthesis* **1992**, 146; *Angew. Chem. Int. Ed. Engl.* **1990**, *29*, 176.

36. Semmelhack, M.F.; Tamura, R. *J. Am. Chem. Soc.* **1983**, *105*, 4099; 6750.

37. Semmelhack, M.F.; Tamura, R.; Schnatter, W.; Springer, J. *J. Am. Chem. Soc.* **1984**, *100*, 5363.

38. Foley, H.C.; Strubinger, L.M.; Targos, T.S.; Geoffroy, G.L. *J. Am. Chem. Soc.* **1983**, *105*, 3064.

39. Geoffroy, G.L. *Adv. Organomet. Chem.* **1985**, *24*, 249.

40. (a) Hegedus, L.S.; DeWeck, G.; D'Andrea, S. *J. Am. Chem. Soc.* **1988**, *110*, 2122. (b) For a review on metal ketene complexes see: Geoffroy, G.L.; Bassner, S.L. *Adv. Organomet. Chem.* **1988**, *28*, 1.

41. (a) Imwinkelried, R.; Hegedus, L.S. *Organometallics* **1988**, *7*, 702. (b) Schwindt, M.A.; Lejon, T.; Hegedus, L.S. *Organometallics* **1990**, *9*, 2814.

42. Hegedus, L.S.; Imwinkelried, R.; Alarid-Sargent, M.; Dvorak, D.; Satoh, Y. *J. Am. Chem. Soc.* **1990**, *112*, 1109.

43. For a discussion of stereoselectivity in these reactions see: Hegedus, L.S.; Montgomery, J.; Narukawa, Y.; Snustad, D.C. *J. Am. Chem. Soc.* **1991**, *113*, 5784.

44. Narukawa, Y.; Juneau, K.N.; Snustad, D.C.; Miller, D.B.; Hegedus, L.S. *J. Org. Chem.* **1992**, *57*, 5453.

45. Valenti, E.; Pericas, M.A.; Moyano, A. *J. Org. Chem.* **1990**, *55*, 3582.

46. Soderberg, B.C.; Hegedus, L.S.; Sierra, M.A. *J. Am. Chem. Soc.* **1990**, *112*, 4364.

47. Soderberg, B.C.; Hegedus, L.S. *J. Org. Chem.* **1991**, *56*, 2209.

48. Aumann, R.; Krüger, C.; Goddard, R. *Chem. Ber.* **1992**, *125*, 1627.

49. Hegedus. L.S.; Bates, R.W.; Soderberg, B.C. *J. Am. Chem. Soc.* **1991**, *113*, 923.

50. Hegedus, L.S.; Schwindt, M.A.; DeLombaert, S.; Imwinkelried, R. *J. Am. Chem. Soc.* **1990**, *112*, 2264.

51. Hegedus, L.S.; Lastra, E.; Narukawa, Y.; Snustad, D.C. *J. Am. Chem. Soc.* **1992**, *114*, 2991.

52. Miller, J.R.; Pulley, S.R.; Hegedus, L.S.; DeLombaert, S. *J. Am. Chem. Soc.* **1992**, *114*, 5602.

53. Merlic, C.A.; Xu, D. *J. Am. Chem. Soc.* **1991**, *113*, 7418.

54. For a review see: Brookhart, M.; Studabaker, W.B. *Chem. Rev.* **1987**, *87*, 411.

55. For examples, see: Fischer, H.; Hoffmann, J. *Chem. Ber.* **1991**, *124*, 981.

56. Mattson, M.N.; Bays, J.P.; Zakutansky, J.; Stolarski, V.; Helquist, P. *J. Org. Chem.* **1989**, *54*, 2467.

57. Vargas, R.M.; Theys, R.D.; Hossain, M.M. *J. Am. Chem. Soc.* **1992**, *114*, 777.

58. Theys, R.D.; Hossain, M.M. *Tetrahedron Lett.* **1992**, *33*, 3447.

59. (a) Brookhart, M.; Liu, Y.; Goldman, E.W.; Timmers, D.A.; Williams, G.D. *J. Am. Chem. Soc.* **1991**, *113*, 927. (b) Brookhart, M.; Liu, Y. *J. Am. Chem. Soc.* **1991**, *113*, 939.

60. For a review see: Adams, J.; Spero, D.M. *Tetrahedron* **1991**, *47*, 1765.

61. (a) Doyle, M.P.; Dorow, R.L.; Buhro, W.E.; Griffin, J.H.; Tamblyn, W.H.; Trudell, M.L. *Organometallics* **1984**, *3*, 44. (b) Doyle, M.P.; Griffin, J.H.; Bagheri, V.; Dorow, R.L. *Organometallics* **1984**, *3*, 53.

62. (a) Davies, H.M.L.; Clark, T.J.; Smith, H.D. *J. Org. Chem.* **1991**, *56*, 3817. (b) Cantrell, W.R., Jr.; Davies, H.M.L. *J. Org. Chem.* **1991**, *56*, 723. (c) Davies, H.M.L.; Clark, T.J.; Kinomer, G.F. *J. Org. Chem.* **1991**, *56*, 6440.

63. Adams, J.; Frenette, R.; Belley, M.; Chibante, F.; Springer, J.P. *J. Am. Chem. Soc.* **1987**, *109*, 5439.

64. Taber, D.F.; Hoerrner, R.S. *J. Org. Chem.* **1992**, *57*, 441.

65. For a review see: Doyle, M.P. *Rec. Trav. Chem.* **1991**, *110*, 305.

66. Doyle, M.P.; Pieters, R.J.; Martin, S.F.; Austin, R.E.; Oalmann, C.J.; Muller, P.J. *J. Am. Chem. Soc.* **1991**, *113*, 1423.

67. (a) Padwa, A.; Austin, D.J.; Xu, S.L. *Tetrahedron Lett.* **1991**, *32*, 4103. (b) Hoye, T.; Dinsmore, C.J. *Tetrahedron Lett.* **1991**, *32*, 3755. (c) Hoye, T.R.; Dinsmore, C.J. *J. Am. Chem. Soc.* **1991**, *113*, 4343. (d) Padwa, A.; Austin, D.J.; Xu, S.L. *J. Org. Chem.* **1992**, *57*, 1330.

68. Edmunds, A.J.F.; Baumann, K.; Grassberger, M.; Schultz, G. *Tetrahedron Lett.* **1991**, *32*, 7039.

69. (a) Lowenthal, R.E.; Masamune, S. *Tetrahedron Lett.* **1991**, *32*, 7373. (b) For related cases see: Evans, D.H.; Woerpel, K.A.; Hinman, M.M.; Faul, M.M. *J. Am. Chem. Soc.* **1991**, *113*, 726. (c) Evans, D.A.; Woerpel, K.A.; Scott. M.J. *Angew. Chem. Int. Ed. Engl.* **1992**, *31*, 430. (d) Leutenegger, U.; Umbricht, G.; Fahrni, C.; von Matt, P.; Pfaltz, A. *Tetrahedron* **1992**, *48*, 2143.

70. Etkin, N.; Babu, S.D.; Fooks, C.J.; Durst, T. *J. Org. Chem.* **1990**, *55*, 1093.

71. (a) Doyle, M.P.; Pieters, R.J.; Taunton, J.; Pho, H.Q.; Padwa, A.; Hertzog, D.L.; Precedo, D.L. *J. Org. Chem.* **1991**, *56*, 820. (b) Wee, A.G.H.; Liu, B.; Zhang, L. *J. Org. Chem.* **1992**, *57*, 4404.

72. Hashimoto, S-I.; Watanabe, N.; Ikegami, S. *Tetrahedron Lett.* **1992**, *33*, 2709. For related directing effects of C–H insertion by ligand alteration see: Spero, D.M.; Adams, J. *Tetrahedron Lett.* **1992**, *33*, 1143.

73. Doyle, M.P.; van Oeveren, A.; Westrum. L.; Protopopova, M.N.; Clayton, T.W., Jr. *J. Am. Chem. Soc.* **1991**, *113*, 8982.

74. McKervey, M.A.; Ye, T. *J. Chem. Soc., Chem. Comm.* **1992**, 823.

75. Ruediger, E.H.; Solomon, C. *J. Org. Chem.* **1991**, *56*, 3183.

76. Heslin, J.C.; Moody, C.J. *J. Chem. Soc., Perkin I* **1988**, 1417.

77. For reviews see: (a) Padwa, A.; Krumpe, K.E. *Tetrahedron* **1992**, *48*, 5385. (b) Padwa, A.; Hornbuckle, S.F. *Chem. Rev.* **1991**, *91*, 263. (c) Padwa, A. *Acct. Chem. Res.* **1991**, *24*, 22.

78. Hertzog, D.L.; Austin, D.J.; Nadler, W.R.; Padwa, A. *Tetrahedron Lett.* **1992**, *33*, 4731.

79. Cannizzo, L.F.; Grubbs, R.H. *J. Org. Chem.* **1985**, *50*, 2386.

80. Cannizzo, L.F.; Grubbs, R.H. *J. Org. Chem.* **1985**, *50*, 2316.

81. For a comparison of Tebbes reagent with Ph_3PCH_2, see: Pine, S.H.; Shen, G.S.; Hoang, H. *Synthesis* **1991**, 165.

82. Philipo, C.M.G.; Vo, N.H.; Paquette, L.A. *J. Am. Chem. Soc.* **1991**, *113*, 2762.

83. Kang, H-J.; Paquette, L.A. *J. Am. Chem. Soc.* **1990**, *112*, 3252.

84. Barrett, A.G.M.; Bezuidenhoudt, B.C.B.; Gasiecki, A.F.; Howell, A.R.; Russell, M.A. *J. Am. Chem. Soc.* **1989**, *111*, 1392.

85. Curtis, N.R.; Holmes, A.B.; Looney, M.G. *Tetrahedron Lett.* **1992**, *33*, 671.

86. Petasis, N.A.; Bzowej, E.I. *J. Am. Chem. Soc.* **1990**, *112*, 6392.

87. DeShong, P.; Rybczynski, P.J. *J. Org. Chem.* **1991**, *56*, 3207.

88. Petasis, N.A.; Bzowej, E.I. *J. Org. Chem.* **1992**, *57*, 1327.

89. Petasis, N.A.; Akritopoulou, I. *Synlett* **1992**, 665.

90. Tour, J.M.; Bedworth, P.V.; Wu, R. *Tetrahedron Lett.* **1989**, *30*, 3927.

91. Buchwald, S.L.; Grubbs, R.H. *J. Am. Chem. Soc.* **1983**, *105*, 5490.

CHAPTER 7

Synthetic Applications of Transition Metal Alkene, Diene, and Dienyl Complexes

7.1   Introduction

Nucleophilic attack on transition metal-complexed olefin, diene, and dienyl systems is among the most useful of organometallic processes for the synthesis of complex organic molecules. Not only does complexation reverse the normal reactivity of these functional groups, changing them from nucleophiles to electrophiles, but the process also results in the formation of new bonds between the nucleophile and the olefinic carbon and new metal-carbon bonds that can be further elaborated (Eq. 7.1). The fundamental features of this process were presented in Chapter 2. Here its uses in organic synthesis will be considered.

7.2   Metal Alkene Complexes

a. Palladium(II) Complexes[1]

Olefin complexes of palladium(II) were among the first used to catalyze useful organic transformations and are among the most extensively developed, at least in part because they are both easily generated and highly reactive. Palladium chloride is the most common catalyst precursor. It is a commercially available red-brown, chlorobridged oligomer, insoluble in most organic solvents (Figure 7.1). However, the oligomer is easily disrupted by treatment with alkali metal chlorides, such as LiCl or NaCl, providing the considerably more soluble, hygroscopic monomeric palladate, M_2PdCl_4. Nitriles, particularly acetonitrile and benzonitrile, also give air-stable, easily handled, soluble monomeric solids, which are the most convenient sources of palladium(II). Treatment of soluble palladium(II) salts with alkenes results in rapid, reversible co-ordination of the alkene to the metal (a in Figure 7.2). Ethylene and ter-minal monoolefins coordinate most strongly followed by *cis* and *trans* internal olefins. Although these olefin complexes can be isolated if desired, they are normally generated *in situ* and used without isolation (Figure 7.2). Once complexed, the olefin becomes *generally* reactive towards nucleophiles ranging from Cl⁻ through Ph⁻, a range of about 10^{36} in basic-ity. Attack occurs from the face opposite the metal (*trans* attack) and at the more substituted carbon, resulting in the formation of a carbon-nucle-ophile bond and a metal-carbon bond (b). Note that this is the opposite regiochemistry and stereochemistry from that obtained by insertion of olefins into metal carbon σ-bonds. The regiochemistry is that correspond-ing to attack at the olefinic position which can *best* stabilize positive charge, and, in a sense, the highly electrophilic palladium(II) is behav-ing like a very expensive and selective proton towards the olefin.

Eq. 7.1

Figure 7.1 Palladium(II) Sources

The newly formed σ-alkylpalladium(II) complex has a very rich chemistry. Warming above –20° results in β-hydrogen elimination (c), generating an olefin which is the product of a formal nucleophilic substitution. This σ-alkylpalladium(II) complex is easily reduced, either by hydrides or hydrogen gas at one atmosphere, resulting in a formal nucleophilic addition to the alkene (d). Carbon monoxide insertion is fast, even at low temperatures, and successfully competes with β-elimination. Thus σ-acyl species are easily produced simply by exposing the reaction solution to an atmosphere of carbon monoxide (e). Cleavage of this acyl species with methanol produces esters (f), while treatment with main group organometallics leads to ketones via a *carbonylative coupling* (g). The initially formed σ-alkylpalladium(II) complex itself will undergo trans-metallation from many main group organometallics, resulting in overall *nucleophilic addition/alkylation* (h).

In principle, olefins can insert into metal-carbon σ-bonds, but in practice this usually cannot be achieved when β-hydrogens are present, since β-elimination is faster. In exceptional cases, however, olefins can insert into σ-alkylpalladium(II) complexes resulting from nucleophilic attack on olefins, resulting in an overall *nucleophilic addition/olefination* (i).

In all of the above processes, palladium(II) is required to activate the alkene, but palladium(0) — from reductive elimination — is the product of the reaction. Thus, for catalysis, a procedure to oxidize Pd(0) back to Pd(II) in the presence of reactants and products is needed. Many systems have been developed ranging from the classic $O_2/CuCl_2$ redox system to more recently developed benzoquinone, or quinone/reoxidant. The choice of reoxidant is usually dictated by the stability of the products toward the oxidant.

Figure 7.2   Reactions of Palladium(II) Olefin Complexes

Palladium(II) olefin complexes are particularly well-behaved in their reactions with oxygen nucleophiles such as water, alcohols, and carboxylates, and a large number of useful synthetic transformations are based on this process. One of the earliest is the Wacker process, the palladium(II) catalyzed "oxidation" of ethylene to acetaldehyde for which the key step is nucleophilic attack of water on metal-coordinated ethylene. Although this particular (industrial) process is of little interest for the synthesis of fine chemicals, when applied to terminal alkenes it provides a very efficient synthesis of methyl ketones (Eq. 7.2).[1c] The reaction is specific for terminal olefins, since they coordinate much more strongly then internal alkenes, and internal alkenes are tolerated elsewhere in the substrate (Eq. 7.3 and Eq. 7.4). As expected, attack always occurs at the 2° carbon, and aldehydes, resulting from attack at the terminal position, are not observed. A variety of remote functional groups are tolerated (Eq. 7.5 and Eq. 7.6) provided they are not strong ligands for palladium(II) (e.g., amines), in which case the catalyst is poisoned. Copper(II) chloride/oxygen is used to reoxidize the palladium, since the product alcohols are stable to these reagents.

Eq. 7.2

Eq. 7.3

59%

Eq. 7.4

78%

Eq. 7.5

87%

Eq. 7.6

85%

This reaction is not useful for internal olefins since they react only slowly, if at all. However, α,β-unsaturated ketones and esters are cleanly converted to β-keto compounds under somewhat specialized conditions (Eq. 7.7).

Eq. 7.7

64%

59%

Alcohols also add cleanly to olefins in the presence of palladium(II) but the intermolecular process has found little use in synthesis. In contrast, the *intra*molecular version provides a very attractive route to oxygen heterocycles. In early examples (Eq. 7.8)[2] and (Eq. 7.9)[3] stoichiometric amounts of palladium(II) salts were required since efficient, mild reoxidants had not yet been developed. In principle all of these cases should be subject to catalysis. In intramolecular reactions, internal and even trisubstituted alkenes undergo attack, and the regioselectivity is often determined more by ring size than by substitution (Eq. 7.10[4] and Eq. 7.11).[5] Because CO insertion into Pd–C bonds is so facile and competes effectively with β-elimination, efficient alkoxycarbonylation of alkenes has been achieved (Eq. 7.12[6] and Eq. 7.13[7]).

Eq. 7.8

85%

Eq. 7.9

30-60%

Eq. 7.10

70-80%

Eq. 7.11

62%   86% ee

Eq. 7.12

Eq. 7.13

Under exceptional circumstances, olefins can be made to insert into σ-alkylpalladium(II) complexes having β-hydrogens. Such a case is seen in Eq. 7.14.[8] The initial step is palladium(II) assisted alcohol attack on the complexed enol ether, generating a σ-alkylpalladium(II) complex, which has a β-hydrogen potentially accessible for β-elimination. However, β-elimination requires a *syn* coplanar relationship and, if the adjacent olefin coordinates to palladium, rotation to achieve this geometry is suppressed. At the same time, coordination of the olefin permits it to insert into the σ-metal carbon bond, which is indeed what occurs, to form the bicyclic system. The insertion is a *syn* process, and the σ-alkylpalladium(II) complex resulting from this insertion cannot achieve a *syn* coplanar geometry with any of the β-hydrogens, since the ring system is rigid. In this special circumstance *inter* molecular olefin insertion occurs, giving the bicyclic product in excellent yield. Note that this process forms one carbon-oxygen and two carbon-carbon bonds in a one pot procedure.

Eq. 7.14

Carboxylate anions are also efficient nucleophiles for palladium assisted attack on olefins. The exceptionally useful palladium(II) catalyzed 1,4-*bis*-acetoxylation or chloroacetoxylation of 1,3-dienes is initiated by *trans* nucleophilic attack of acetate on the palladium-complexed diene (Eq. 7.15),[9] generating a π-allylpalladium(II) complex. Since it is the reactions of this π-allyl complex with nucleophiles which lead to the synthetically interesting products, this topic is discussed in Chapter 9.

Intramolecular versions are efficient (Eq. 7.16) and, as is often the case for intramolecular reactions, even trisubstituted olefins are attacked.[10]

Eq. 7.15

ox = Benzoquinone/MnO₂ or
O₂/ Hydroquinone/Co (salophen)

ox = $Benzoquinone/MnO_2$ or
O_2/ Hydroquinone/Co (salophen)

Eq. 7.16

55-80%

Amines also attack palladium-bound olefins, but catalytic inter-molecular amination of olefins has not been achieved, since both the amine and the product enamine are potent ligands for palladium and act as catalyst poisons. The stoichiometric amination of simple olefins can be achieved in good yield, but three equivalents of amine are required, and the resulting σ-alkyl palladium(II) complex must be reduced to give the free amine.[11] A careful study of intramolecular amination elucidated some of the difficulties (Eq. 7.17).[12] Treatment of the cyclopentyl aminoolefin with palladium(II) chloride resulted in the formation of a stable, chelated aminoolefin complex, confirming that one problem is indeed the strong coordination of amines to Pd(II). N-Acylation of the amine dramatically decreased its basicity and coordinating ability, and this acetamide did not irreversibly bind to palladium(II). Instead intramolecular amination of the olefin ensued. However, the resulting σ-alkyl palladium(II) complex was stabilized by chelation to the amide oxygen, preventing β-hydrogen elimination, and thus catalysis. Finally, by placing the strongly electron withdrawing sulfonyl group on nitrogen, neither the substrate nor the product coordinated strongly enough to palladium(II) to prevent catalysis, and efficient cyclization was achieved.

Eq. 7.17

Aromatic amines are about 10^6 less basic than aliphatic amines, and this factor permits the efficient intramolecular amination of olefins by these amines without the need for N-acylation or sulfonation, although these N-functionalized substrates also cyclize efficiently (Eq. 7.18).[13] A variety of functionalized indoles can be made by this procedure.

Eq. 7.18

The σ-alkylpalladium(II) complex from the amination of olefins can easily be trapped by CO (Eq. 7.19[14] and Eq. 7.20[15]). With careful choice of conditions even olefins will insert reasonably efficiently although β-hydride elimination competes even under the best of circumstances (Eq. 7.21).[16] Note that although there are nine sites at which the catalyst can (and almost certainly does) coordinate, only one leads to cyclization. Provided that the catalyst is not irreversibly bound to one of the other sites, catalysis ensues. Palladium-catalyzed amination of olefins in conjunction with oxidative addition/insertion chemistry (Chapter 4) has resulted in a very efficient synthesis of claviciptic acid methyl ester (Eq. 7.22).[17]

Eq. 7.19

Eq. 7.20

Eq. 7.21

Eq. 7.22

(18% overall yield from commercial
2-bromo-6-nitrotoluene
vs. 0.4% by classical methods)

The use of carbon nucleophiles results in yet another set of problems. Carbanions, even stabilized ones, are easily oxidized, and palladium(II) salts, as well as many other transition metal salts, tend to oxidatively dimerize carbanions, preventing their attack on olefins. Even if this can be prevented, catalysis is (as yet) impossible to achieve since it is necessary to find an oxidizing agent capable of reoxidizing the catalytic amount of Pd(0) present to Pd(II), but at the same time, not able to oxidize the large amount of the carbanion substrate present. Stoichiometric alkylation of olefins by stabilized carbanions can be achieved by very careful choice of conditions (Eq. 7.23).[18] Preforming the olefin complex at 0° followed by cooling to –78° and addition of two equivalents of triethyl amine generates an olefin/amine complex. The coordinated amine directs the carbanion away from the metal and to the olefin, suppressing metal centered electron transfer redox chemistry, and promoting alkylation of the olefin. As usual, attack occurs at the more substituted carbon of the olefin, producing a σ-alkylpalladium(II) complex which can undergo reduction, β-elimination, or CO insertion. The Pd(0) produced can be recovered and reoxidized separately, but not concurrently with alkylation, so the process is stoichiometric in palladium, and unlikely to find extensive use in synthesis.

Eq. 7.23

R = H, Ph, Me, nBu, AcNH
X, Y = CO₂R, COR, CN

The exception to this will be cases in which this chemistry permits the efficient synthesis of products which would be considerably more difficult to prepare by classical means. Such a situation is seen in the synthesis of a relay to (+)-thienamycin (Eq. 7.24).[19] In this case, one pot alkylation/acylation of an optically active ene carbamate produced the keto diester in 70% chemical yield, and with complete control of stereochemistry at the chiral center α- to nitrogen. In this case, palladium was used to form two C–C and one O–C bond, and to generate a new chiral center. When palladium(II) assisted alkylation of this olefin was combined with carbonylative coupling to tin reagents, even more complex structures were efficiently produced (Eq. 7.25).[20] In this case, palladium(II) promotes stereospecific alkylation of the olefin, followed by insertion of CO into the metal carbon σ-bond, followed by transmetallation from tin to palladium, followed by reductive elimination. In this case three C–C bonds and a chiral center are efficiently generated in a one-pot reaction.

Eq. 7.24

Eq. 7.25

60-95%
≥97% de

Palladium(II) acetate also promotes the intramolecular alkylation of olefins by silylenol ethers (Eq. 7.26[21] and Eq. 7.27[22]). It should be possible to make this process proceed catalytically since silylenol ethers themselves should not be oxidatively dimerized by Pd(II). However, some technical problem exists, since only one turn on Pd(II) has been achieved.

Eq. 7.26

80%

Eq. 7.27

55%

Palladium(II) complexes catalyze the rearrangement of a very broad array of allylic systems, in a process which involves intramolecular nucleophilic attack on a Pd(II)-complexed alkene as the key step (Eq. 7.28).[23] Allyl acetates rearrange under mild conditions without complication from skeletal rearrangements or formation other side products (Eq. 7.29 and Eq. 7.30). With optically active allyl acetates, rearrangement occurs with complete transfer of chirality (Eq. 7.31).[24]

Eq. 7.28

Eq. 7.29

Eq. 7.30

Eq. 7.31

R = CH₃, OBn

Palladium(II) complexes catalyze the Cope rearrangement of 1,5-dienes under very mild conditions, leading to roughly 10^{10} rate enhancement over the thermal process (Eq. 7.32).[23] There are some limitations to the catalyzed process, in that C-2 must bear a substituent to help stabilize the developing positive charge, and C-5 must bear a hydrogen so that efficient complexation to Pd(II) results. (Recall that trisubstituted olefins complex poorly to Pd(II), and that *gem* disubstituted alkenes undergo attack at the substituted position, making Cope rearrangements disfavored for these systems under Pd(II) catalysis.) Again, with chiral substrates, complete chirality transfer is observed.[25] Palladium(II) complexes also catalyze the oxyCope rearrangement (Eq. 7.33).[26]

Eq. 7.32

Eq. 7.33

Many other allylic systems undergo this palladium(II) catalyzed allylic transposition. Trichloromethylimidates rearrange cleanly, and with complete transfer of chirality (Eq. 7.34[27] and Eq. 7.35[28]). O-Allyl oximes rearrange to nitrones, which can be efficiently utilized in 1,3-dipolar cycloaddition reactions (Eq. 7.36).[29]

Eq. 7.34

Eq. 7.35

Eq. 7.36

b. Iron(II) Complexes

Cationic olefin complexes of iron(II) are easily made by alkene exchange with the stable $CpFe(CO)_2$(isobutene)$^+$ complex which is a stable gold-yellow solid, made by the reaction of Fp$^-$ (Fp = $CpFe(CO)_2$) with methallyl bromide, followed by γ-protonation with HBF_4 (see Chapter 2) (Eq. 7.37).[30] Once complexed, the olefin becomes reactive towards nucleophilic attack, sharing most of the reactivity, regio- and stereochemical features of the neutral olefin palladium(II) complexes discussed above. However, in contrast to the palladium analogs, the resulting σ-alkyl-iron(II) complex is very stable, and iron must be removed in a separate chemical step, usually oxidation. As an added complication, oxidative cleavage of Fp alkyl complexes usually results in insertion of CO into the metal-carbon σ-bond, producing carbonylated products (Eq. 7.38). For these reasons, iron-olefin complexes have been used in synthesis considerably less often than those of palladium. However, the reaction chemistry of electron rich alkenes complexed to iron is quite unusual, and synthetically useful.

Eq. 7.37

Eq. 7.38

stable, gold-yellow
solid

When complexed to the Fp cation, dimethoxyethene is quite reactive towards nucleophiles, and both methoxy groups can be replaced, making this iron olefin complex essentially a vinylidene dication (CH^+=CH^+) equivalent.[31] Treatment with optically active 2,3-butandiol generates the Fp$^+$-dioxene complex in good yield (Eq. 7.39). The optically active dioxene is easily freed by treatment with iodide, providing a simple route to optically active compounds not easily prepared. More interestingly, ketone enolates attack the dioxene complex from the face opposite the metal to give the stable Fp alkyl complex with high diastereoselectivity. Stereoselective reduction of the carbonyl group with L-Selectride, followed by oxidative removal of the metal produces the tricyclic furan in excellent yield and with high diastereoselectivity. In this case oxidation of the iron does not result in CO insertion, but rather it simply makes the iron a good leaving group (Eq. 7.39).[32]

Eq. 7.39

An unusual class of reactions involving cationic iron olefin complexes is the "3+2" cycloaddition reaction that occurs between electron poor alkenes and η^1-allyl Fp complexes (Eq. 7.40).[33] In this case, the η^1-allyl Fp complex nucleophilically attacks the electrophilic alkene, producing a stabilized enolate and a cationic Fp olefin complex. The enolate then attacks the cationic olefin complex, closing the ring, and generating a Fp alkyl complex. Oxidative removal of the iron promotes CO insertion, and generates the highly functionalized cyclopentane ring system.

Eq. 7.40

$R^1, R^2 = CN, CO_2R$
$R^3, R^4 = H, CO_2R, CN$

The same chemical features that make cationic iron-olefin complexes reactive towards nucleophiles makes them unreactive towards electrophiles and complexation of alkenes to Fp^+ can protect them against an array of electrophilic reactions. Complexation occurs preferentially at the less substituted alkene, and at alkenes rather than alkynes, so selective protection is possible (Eq. 7.41, Eq. 7.42, and Eq. 7.43).[34] The olefin is freed of iron after reaction by treatment with iodide.

Eq. 7.41

Eq. 7.42

Eq. 7.43

7.3 Metal Diene Complexes

a. Fe(CO)$_3$ as a 1,3-Diene Protecting Group

Heating Fe$_2$(CO)$_9$ with dienes produces iron carbonyl complexes of conjugated diene (Eq. 7.44). Non-conjugated dienes often rearrange to give conjugate diene complexes in the presence of Fe$_2$(CO)$_9$ or Fe(CO)$_5$. These iron tricarbonyl complexes of dienes are very stable: they fail to undergo Diels-Alder reactions, and undergo facile Friedel-Crafts acylation without decomplexation from the metal. Although they can be made to undergo reaction with strong nucleophiles (see next section), complexation of the diene segment is most often used to protect it.

Eq. 7.44

For example, dienals are highly reactive and polymerize readily, making reactions at the aldehyde difficult and inefficient. However, treatment with Fe$_2$(CO)$_9$ forms the stable diene complex, in which the aldehyde reacts normally (Eq. 7.45,[35] Eq. 7.46,[36] and Eq. 7.47[37]). When the reactions are complete, the iron is oxidatively removed, giving the metal-free organic compound. Particularly notable is the range of reactions the iron tricarbonyl group will withstand. These include Wittig, Aldol, reduction, and even the *cis* hydroxylation of an olefin with osmium tetroxide.

Eq. 7.45

Eq. 7.46

Eq. 7.47

Because iron occupies a single face of these diene complexes, substituted diene complexes are intrinsically chiral, and the presence of the iron has a substantial stereochemical influence. For example reaction of racemic iron diene complex in Eq. 7.48 with S,S allylboronates resulted in efficient kinetic resolution of the iron complex giving 96% ee at 60% conversion. In this case the 2R enantiomer of the iron complex underwent reaction much faster than the 2S enantiomer. When resolved 2R complex was used, very high (>100:1) stereoselectivity at the secondary alcohol formed was observed (Eq. 7.48).[38]

Eq. 7.48

A related example of "double diastereoselection"[39] was observed in the reaction of the optically active, iron-complexed ylide with the optically active epoxy aldehyde in Eq. 7.49.[40] With racemic ylide, a diastereoselectivity of 4:1 was observed, with the 2S,5R iron/5S,6S aldehyde being the "matched" pair and the 2R,5S iron/5S,6S aldehyde being the mismatched pair. With resolved iron complexes, the mismatched pair reacted very slowly and in low yield, but, because all existing stereogenic centers were fixed, and no new stereogenic centers were generated, the reaction was still stereospecific.

Complexation of dienes to iron tricarbonyl fragments can also be used to both activate and stabilize allylic positions, as well as to exert stereochemical control. The methyl group in complexed isoprene was easily deprotonated, then alkylated by organic halides (Eq. 7.50).[41] When a new chiral center was formed in this alkylation, very high stereoselectivity, relative to iron, could be achieved (Eq. 7.51).[42] Complexation of the cyclohepta-3,5-dienone permitted α-dialkylation with clean *cis* stereochemistry (Eq. 7.52).[43] In this case, the iron fragment prevented isomerization of the enolate and directed the alkyl groups to the face opposite the metal. Reduction of the resulting α,α'-dimethylketone also occurred stereospecifically, but, because of the flanking methyl groups, attack occurred from the same face as the metal. Similarly, two of the three conjugated double bonds of cyclohepta-2,4,6-trien-1-ol could be complexed to iron, permitting oxidation or hydroboration of the uncomplexed olefin with very high stereoselectivity (Eq. 7.53).[44]

Eq. 7.49

Eq. 7.50

Eq. 7.51

R = Me, Bn,

90%, >97% de when R = Me

Eq. 7.52

Eq. 7.53

Similar activation and directing effects have been achieved using complexation of dienes to molybdenum(II) (Eq. 7.54).[45] In this case, the cationic diene complex was converted to the neutral π-allylmolybdenum complex by allylic proton abstraction. This protected one of the double bonds of the diene, permitting functionalization of the other. Cycloheptadiene underwent a similar series of reactions.

Eq. 7.54

Diene complexes are also produced by the reaction of 1,4-dihalo-2-butenes with $Fe_2(CO)_9$ (Eq. 7.55). An especially useful example of this reaction is the preparation of (η^4-cyclobutadiene)iron tricarbonyl from dichlorocyclobutene[46] (Eq. 7.56). This reaction illustrates the stabilization of extremely unstable organic molecules by complexation to appropriate transition metals. Free cyclobutadiene itself has only fleeting existence, since it rapidly dimerizes. However, cyclobutadieneiron tricarbonyl is a very stable complex. The complexed cyclobutadiene undergoes a variety of electrophilic substitution reactions, including Friedel Crafts acylation, formylation, chloromethylation, and aminomethylation. The keto group of the acylated cyclobutadiene iron tricarbonyl complex can be reduced by hydride reagents without decomposition.[47] On the other hand, oxidation of the complex with cerium(IV), triethylamine oxide, or pyridine N-oxide frees the cyclobutadiene ligand for use in synthesis. The use of cyclobutadiene iron tricarbonyl as a stable source of highly reactive cyclobutadiene has been exploited for the synthesis of cubane.[48]

Eq. 7.55

Eq. 7.56

b. Nucleophilic Attack on Metal-Diene Complexes

Although complexation of dienes to electron-deficient metal fragments stabilizes these ligands towards electrophilic attack, it activates them towards nucleophilic attack, a reaction mode unavailable to the uncomplexed ligand. Neutral η^4-1,3-diene iron tricarbonyl complexes are reactive only towards very strong nucleophiles, and attack can occur at either the terminal or internal position of the diene. Kinetically, attack at the terminal position to form an η^3-allyliron complex is favored. However, this is a reversible process, and thermodynamically, attack at an internal position generating a η^1-alkyl-η^2-olefin iron complex is favored (Eq. 7.57).[49] As usual, attack occurs from the face opposite the metal.

Eq. 7.57

$R^- = CMe_2CN, CHMeCN$
CH_2CN, CMe_2CO_2Et
CMe_2CO_2Li
Ph_2CH_2

Protolytic cleavage of the "thermodynamic" iron complex gives the alkylation product, while treatment with carbon monoxide results in insertion into the iron-carbon σ-bond. Protolytic cleavage of this σ-acyl complex produces the *trans* aldehyde.

With acyclic dienes, the situation is a little more complicated. Complexed butadiene itself is converted cleanly to the corresponding cyclopentanone, by a process (Eq. 7.58) involving nucleophilic attack (a), followed by CO insertion (b), followed by olefin insertion (c). However, substitution on the diene suppresses the olefin insertion step, and the amount of the "normal" cleavage product (d) (eg., aldehyde) increases as substitution increases.[50] This olefin-insertion followed by migration of iron to the position α to the carbonyl group is precedented in the insertion of ethylene into iron complexes generated from $Fe(CO)_4^{2-}$. In the absence of CO, the σ-alkyl iron complex will undergo reaction with a range of electrophiles, to result in overall 1,2-difunctionalization of the diene (Eq. 7.59).[51]

Eq. 7.58

Eq. 7.59

By making the metal fragment more electrophilic, the reactivity of metal diene complexes toward nucleophiles can be increased. This is often achieved by replacing a neutral metal fragment by a cationic one, such as the $CpMo(CO)_2^+$ fragment. These complexes are relatively easy to synthesize, although several steps are required (Eq. 7.60).[52] They undergo reaction with a much wider range of nucleophiles, exclusively at the terminal position, to generate a neutral η^3-allylmolybdenum complex. Hydride abstraction from this complex regenerates a cationic diene complex which again undergoes nucleophilic attack from the face opposite the metal, resulting in overall *cis*-1,4 difunctionalization of the diene (Eq. 7.61).[53] Cyclohepta-1,3-diene undergoes a similar series of transformations. The major limitation, to date, is finding an efficient way to remove the molybdenum from the final η^3-allyl complex. Modest success has been achieved by conversion of the neutral η^3-allyl molybdenum carbonyl complex to the cationic nitrosyl (exchange of neutral CO for NO$^+$) permitting nucleophilic attack on the η^3-allyl complex (Eq. 7.62).[54] However the yields for this process are only modest. Of particular synthetic interest is the application of this methodology to heterocyclic dienes (Eq. 7.63).[55]

Eq. 7.60

Eq. 7.61

Eq. 7.62

(R =CH₂CH₂COOH)

R = (CH₂)₄CO₂Et, (CH₂)₃CO₂Me,
(CH₂)₂CO₂Me, PhCH₂, AcO(CH₂)₄,
NC(CH₂)₃, NC(CH₂)₄

35%

Eq. 7.63

60-90%

92%

92%

(1) CF₃COOH

(2) H₂/Pd
72%

R, R′ = D, Me, *p*tol, RC≡C, CH₂CO₂*t*Bu,

c. Nucleophilic Attack on Cationic Metal Dienyl Complexes[56]

Cationic *dienyl* complexes of iron are among the most extensively used complexes in organic synthesis. They are generally reactive towards a wide range of nucleophiles, and the presence of iron permits a high degree of both stereo- and regiocontrol. Cyclohexadienyl complexes are the most studied, at least in part because a wide array of substituted cyclohexadienes are readily available from the Birch reduction of aromatic compounds. The cyclohexadienyl iron complexes are prepared by simply treating either the conjugated or nonconjugated diene with $Fe_2(CO)_9$ (Eq. 7.64). Treatment of the resulting neutral diene complex with trityl cation (hydride abstraction) produces the relatively air- and moisture-stable cationic cyclohexadienyl iron complex. A very wide range of nucleophiles attack this complex at the terminus of the dienyl system from the face opposite the metal, regenerating the neutral diene complex. Treatment with trityl cation removes hydride from the unsubstituted position of the diene complex, and the resulting dienyl complex again reacts with nucleophiles at this position from the face opposite the metal, giving a *cis*-1,2-disubstituted cyclohexadiene. The iron is easily removed by treatment with amine oxides, giving the free organic compound (Eq. 7.61).

Eq. 7.64

Substituted complexes can be prepared and undergo reaction with high regioselectivity. The diene complex derived from 4-methyl anisole undergoes hydride abstraction α to the methoxy group, giving a single regioisomer of the dienyl complex. Nucleophilic attack occurs exclusively at the *methyl* terminus of the dienyl system, apparently because the other end is electronically deactivated by the strongly electron-donating methoxy group. Oxidative removal of the iron followed by hydrolysis gives the 4,4-disubstituted cyclohexenone (Eq. 7.65[57] and Eq. 7.66[58]).

Eq. 7.65

Eq. 7.66

Cationic dienyliron complexes are potent electrophiles, and even undergo electrophilic aromatic substitution reactions with electron rich arenes (Eq. 7.67[59] and Eq. 7.68[60,61]).

Eq. 7.67

Eq. 7.68

An example of the utility of this (and other) organometallic reaction chemistry in the synthesis of complex molecules is seen in Eq. 7.69.[62]

Because iron tricarbonyl occupies a single face of the η^5-dienyl complex, unsymmetrically substituted complexes are intrinsically chiral and can be resolved (Eq. 7.70).[63] Since addition of nucleophiles is stereospecific, these complexes are of use in asymmetric synthesis.

Eq. 7.69

Eq. 7.70

mixture of
diastereoisomers

(+)-single enantiomer

The above reaction chemistry is not restricted to cyclohexadienes. The same degree of activation, regio- and stereoselectivity is achieved with cycloheptadiene.[64] In contrast, acyclic η^5-dienyl complexes have been much less studied, and have only rarely been used in synthesis. They are more difficult to prepare, and often react with lower regioselectivity than do their cyclic analogs. However, once formed they are generally reactive towards nucleophiles (Eq. 7.71[65] and Eq. 7.72[66]).

Eq. 7.71

33-93%

Nuc = H_2O, H^-, Ph_3P,

Eq. 7.72

27-64%

References

1. (a) Heck, R.F. *Palladium Reagents in Organic Synthesis*. Academic Press, New York, 1985. (b) Tsuji, J. *Organic Synthesis with Palladium Compounds*. Springer-Verlag, Berlin, 1980. (c) Tsuji, J. *Synthesis* **1984**, 369.
2. Pearlman, B.A.; McNamara, J.M.; Hasan, I.; Hatakeyama, S.; Sekizaki, H.; Kishi, Y. *J. Am. Chem. Soc.* **1981**, *103*, 4248.
3. Hosokawa, T.; Murahashi, S-I. *Acct. Chem. Res.* **1990**, *23*, 49.
4. Kasahara, A.; Izumi, T.; Ooshima, M. *Bull. Chem. Soc. Jpn.* **1974**, *47*, 2526.
5. Saito, S.; Hara, T.; Takahashi, N.; Hirai, M.; Moriwake, T. *Synlett* **1992**, 237.
6. Semmelhack, M.F.; Zask, A. *J. Am. Chem. Soc.* **1983**, *105*, 2034.
7. (a) Gracza, T.; Hasenöhnl, T.; Stahl, U.; Jäger, V. *Synthesis* **1991**, 1108. (b) Gracza, T.; Jäger, V. *Synlett* **1992**, 191.
8. Larock, R.C.; Lee, N.H. *J. Am. Chem. Soc.* **1991**, *113*, 7815.
9. (a) Backvall, J.E.; Bystrom, S.E.; Nordberg, R.E. *J. Org. Chem.* **1984**, *49*, 4619. (b) Backvall, J.E.; Hopkins, R.B.; Grennberg, H.; Mader, M.M.; Awasthi, A.K. *J. Am. Chem. Soc.* **1990**, *112*, 5160. (c) Grennberg, H.; Gogoll, A.; Backvall, J.E. *J. Org. Chem.* **1991**, *56*, 5808.
10. Korte, D.E.; Hegedus, L.S.; Wirth, R.K. *J. Org. Chem.* **1977**, *42*, 1329.
11. (a) Akermark, B.; Backvall, J-E.; Hegedus, L.S.; Zetterberg, K.; Siirala-Hansen, K.; Sjoberg, K. *J. Organomet. Chem.* **1974**, *72*, 127. (b) Akermark, B.; Akermark, G.; Hegedus, L.S.; Zetterberg, K. *J. Am. Chem. Soc.* **1981**, *103*, 3037.
12. Hegedus, L.S.; McKearin, J.M. *J. Am. Chem. Soc.* **1982**, *104*, 2444.
13. For a review see: Hegedus, L.S. *Angew. Chem. Int. Ed. Engl.* **1988**, *27*, 1113.

14. Tamaru, Y.; Tanigawa, H.; Itoh, S.; Kimura, M.; Tanaka, S.; Fugami, K.; Sekiyama, T.; Yoshida, Z-I. *Tetrahedron Lett.* **1992**, *33*, 631.
15. Gallagher, T.; Davies, I.W.; Jones, S.W.; Lathbury, D.; Mahon, M.F.; Molloy, K.C.; Shaw, R.W.; Vernon, P. *J. Chem. Soc. Perkin I* **1992**, 433.
16. Weider, P.R.; Hegedus, L.S.; Asada, H.; D'Andreq, S.V. *J. Org. Chem.* **1985**, *50*, 4276.
17. Harrington, P.J.; Hegedus, L.S.; McDaniel, K.F. *J. Am. Chem. Soc.* **1987**, *109*, 4335.
18. (a) Hegedus, L.S.; Williams, R.E.; McGuire, M.A.; Hayashi, T. *J. Am. Chem. Soc.* **1980**, *102*, 4973. (b) Hegedus, L.S.; Darlington, W.H. *J. Am. Chem. Soc.* **1980**, *102*, 4980.
19. Montgomery, J.; Wieber, G.M.; Hegedus, L.S. *J. Am. Chem. Soc.* **1990**, *112*, 6255.
20. Masters, J.J.; Hegedus, L.S.; Tamariz, J. *J. Org. Chem.* **1991**, *56*, 5666.
21. Kende, A.S.; Roth, B.; Sanfilippo, P.J. Blacklock, T.J. *J. Am. Chem. Soc.* **1982**, *104*, 5808.
22. Kende, A.S.; Roth, B.; Sanfilippo, P.J.; *J. Am. Chem. Soc.* **1982**, *104*, 1784.
23. For a review see: Overman, L.E. *Angew. Chem. Int. Ed. Engl.* **1984**, *23*, 579.
24. (a) Grieco, P.A.; Takigawa, T.; Bongers, S.L.; Tanaka, H. *J. Am. Chem. Soc.* **1980**, *102*, 7587. (b) Grieco, P.A.; Tuthill, P.A.; Sham, H.L. *J. Org. Chem.* **1981**, *46*, 5005.
25. Overman, L.E.; Jacobsen, E.J. *J. Am. Chem. Soc.* **1982**, *104*, 7225.
26. Bluthe, N.; Malacria, M.; Gore, J. *Tetrahedron Lett.* **1983**, *24*, 1157.
27. Mehmandoust, M.; Petit, Y.; Larcheveque, M. *Tetrahedron Lett.* **1992**, *33*, 4313.
28. Metz, P.; Mues, C.; Schoop, A. *Tetrahedron* **1992**, *48*, 1071.
29. Grigg, R.; Markandu, J. *Tetrahedron Lett.* **1991**, *32*, 279.
30. Cutler, A.; Ehntholt, D.; Lennon, P.; Nicholas, K.; Marten, D.F.; Madhavarao, M.; Raghu, S.; Rosan, A.; Rosenblum, M. *J. Am. Chem. Soc.* **1975**, *97*, 3149.
31. Marsi, M.; Rosenblum, M. *J. Am. Chem. Soc.* **1984**, *106*, 7264.
32. Rosenblum, M.; Foxman, B.M.; Turnbull, M.M. *Heterocycles* **1987**, *25*, 419.
33. For a review see: Welker, M.E. *Chem. Rev.* **1992**, *92*, 97.
34. (a) Boyle, P.F.; Nicholas, K.M. *J. Org. Chem.* **1975**, *40*, 2682. (b) Nicholas, K.M. *J. Am. Chem. Soc.* **1975**, *97*, 3254.
35. Franck-Neumann, M.; Martina, D.; Heitz, M.P. *Tetrahedron Lett.* **1982**, *23*, 3493.
36. Donaldson, W.A.; Tao, C.; Bennett, D.W.; Grubisha, D.S. *J. Org. Chem.* **1991**, *56*, 4563.
37. Gigou, A.; Beaucourt, J-P.; Lellouche, J-P.; Gree, R. *Tetrahedron Lett.* **1991**, *32*, 635.
38. Roush W.R.; Park, J.C. *Tetrahedron Lett.* **1990**, *31*, 4707.
39. For a review see: Masamune, S.; Choy, W.; Peterson, J.S.; Sita, L.R. *Angew. Chem. Int. Ed. Engl.* **1985**, *24*, 1.
40. Pinsard, P.; Lellouche, J-P.; Beaucourt, J-P.; Gree, R. *Tetrahedron Lett.* **1990**, *31*, 1141.
41. Semmelhack, M.F.; Fewkes, E.J. *Tetrahedron Lett.* **1987**, *28*, 1497.
42. Donaldson, W.A.; Craig, R.; Spanton, S. *Tetrahedron Lett.* **1992**, *33*, 3967.
43. Pearson, A.J.; Chang, K. *J. Chem. Soc., Chem. Comm.* **1991**, 394.
44. Pearson, A.J.; Srinivasan, K. *J. Chem. Soc., Chem. Comm.* **1991**, 392.

45. Pearson, A.J.; Mallik, S.; Mortezaei, R.; Perry, M.W.D.; Shively, R.J., Jr.; Youngs, W.J. *J. Am. Chem. Soc.* **1990**, *112*, 8034.
46. Emerson, G.F.; Watts, L.; Pettit, R. *J. Am. Chem. Soc.* **1965**, *87*, 131.
47. Fitzpatrick, J.D.; Watts, L.; Emerson, G.F.; Pettit, R. *J. Am. Chem. Soc.* **1965**, *87*, 3254.
48. Barbarak, J.C.; Watts, L.; Pettit, R. *J. Am. Chem. Soc.* **1966**, *88*, 1328.
49. (a) Semmelhack, M.F.; Herndon, J.W. *Organometallics* **1983**, *2*, 363. (b) Semmelhack, M.F.; Herndon, J.W.; Springer, J.P. *J. Am. Chem. Soc.* **1983**, *105*, 2497.
50. (a) Semmelhack, M.F.; Herndon, J.W.; Liu, J.K. *Organometallics* **1983**, *2*, 1885. (b) Semmelhack, M.F.; Le, H.T.M. *J. Am. Chem. Soc.* **1985**, *107*, 1455.
51. Yeh, M.C.P.; Hwu, C.C. *J. Organomet. Chem.* **1991**, *419*, 341.
52. Faller, J.W.; Murray, H.H.; White, D.L.; Chao, K.H. *Organometallics* **1983**, *2*, 400.
53. (a) Pearson, A.J.; Khan, M.N.I.; Clardy, J.C.; Ciu-Heng, H. *J. Am. Chem. Soc.* **1985**, *107*, 2748. (b) Pearson, A.J.; Khan, M.N.I. *Tetrahedron Lett.* **1984**, *25*, 3507. (c) Pearson, A.J.; Khan, M.N.I. *Tetrahedron Lett.* **1985**, *26*, 1407.
54. Yeh, M.C.P.; Tsou, C-J.; Chuang, C-N.; Lin, H-W. *J. Chem. Soc., Chem. Comm.* **1992**, 890.
55. Hansson, S.; Miller, J.F.; Liebeskind, L.S. *J. Am. Chem. Soc.* **1990**, *112*, 9660.
56. For a review see: Pearson, A.J. "Comprehensive Organic Synthesis", **1991**, *4*, 663.
57. (a) Pearson, A.J. *Acc. Chem. Res.* **1980**, *13*, 463. (b) Pearson, A.J.; Richards, I.C. *Tetrahedron Lett.* **1983**, *24*, 2465.
58. Potter, G.A.; McCague, R. *J. Chem. Soc., Chem. Comm.* **1992**, 635.
59. Stephenson, G.R. *J. Organomet. Chem.* **1985**, *286*, C41.
60. (a) Knölker, H.J.; Bauermeister, M.; Blaser, D.; Boese, R.; Pannek, J-B. *Angew. Chem. Int. Ed. Engl.* **1989**, *28*, 223. (b) Knölker, H.J.; Bauermeister, M. *J. Chem. Soc., Chem. Comm.* **1990**, 664.
61. For a review on the use of this chemistry in the synthesis of heterocycles see: Knölker, H.J. *Synlett* **1992**, 371.
62. Pearson, A.J.; Ray, T. *Tetrahedron Lett.* **1985**, *26*, 2981.
63. (a) Bandara, B.M.R.; Birch, A.J.; Kelley, L.F.; Khor, T.C. *Tetrahedron Lett.* **1983**, *24*, 2491. (b) Howell, J.A.S.; Thomas, M.J. *J. Chem. Soc. Dalton Trans* **1983**, 1401. Atton, J.G.; Evans, D.J.; Kane-Maguire, L.A.P.; Stephenson, G.R. *J. Chem. Soc. Chem. Comm.* **1984**, 1246.
64. (a) Pearson, A.J.; Kole, S.L.; Yoon, J. *Organometallics* **1986**, *5*, 2075. (b) Pearson, A.J.; Ray, T. *Tetrahedron Lett.* **1986**, *27*, 3111.
65. Donaldson, W.A. *J. Organomet. Chem.* **1990**, *395*, 187.
66. Laabassi, M.; Gree, R. *Bull Soc. Chim. Fr.* **1992**, *129*, 151.

CHAPTER 8

Synthetic Applications of Transition Metal Alkyne Complexes

8.1 Introduction

Although virtually all transition metals react with alkynes, relatively few form simple, stable metal alkyne complexes analogous to the corresponding metal olefin complexes. This is because many alkyne complexes are quite reactive toward additional alkyne, and react further to produce more elaborate complexes or organic products. Those that are stable tend to be so stable that use in synthesis is precluded by their lack of reactivity. The successful development of transition-metal-alkyne complex chemistry for use in organic synthesis has involved procedures for the management of these extremes in reactivity.

8.2 Nucleophilic Attack on Metal Alkyne Complexes

In contrast to metal-olefin complexes nucleophilic attack on metal-alkyne complexes is relatively rare and has found little use in synthesis. Stable $CpFe(CO)_2^+$ alkyne complexes undergo attack by a wide range of nucleophiles, forming very stable σ-vinyliron complexes, which can be made to insert CO by treatment with $Ag^{(+)}$ (Lewis Acid promoted insertion) (Eq. 8.1).[1] Although these complexes should have additional reactivity, little has been done with them.

Eq. 8.1

$$Nuc^- = R'_2CuLi, PhS^-, (-) \diagdown \begin{smallmatrix} Y \\ X \end{smallmatrix}$$

Palladium(II) complexes catalyze the cyclization of homopropargyl alcohols and amines, in a process that probably involves nucleophilic attack on a complexed alkyne (Eq. 8.2)[2] similar to that observed with palladium(II) complexed alkenes (Chapter 7). However, the mechanism of this process has not been studied, and the putative protolytic cleavage of the σ-vinylpalladium(II) complex by the protonated heterocycle is un-usual. (The formation of the five-membered ring by the *cis* alkynol and the six-membered ring by the *trans* alkynol is likely due to the strain associated with *trans* [3.3.0] ring systems.)

Eq. 8.2

Palladium(0) complexes also catalyze a series of "alkylative" het-erocyclizations of alkynes (Eq. 8.3[3] and Eq. 8.4[4]). However, these probably involve oxidative addition/insertion chemistry rather than nucleophilic attack on the alkyne, and are most closely related to the "Heck" reaction discussed in Chapter 7.

Eq. 8.3

R^1 = H, Me, Ac, Ts

R^2 = nPr, tBu, C_6H_{11}, , TMS, Ph, CH_2OH

R^3 = nPr, Me, Et, , CH_2OH, Ph

Eq. 8.4

via $R'X$ + Pd(0) ⟶ R'—Pd—X $\xrightarrow[\text{insert}]{\text{substrate}}$

8.3 Stable Alkyne Complexes

a. As Alkyne Protecting Groups

In contrast to most other metals, dicobalt octacarbonyl forms very stable complexes with alkynes, which act as four electron donor, bridging ligands, perpendicular to the Co–Co bond (Eq. 8.5).[5] This complexation effectively reduces the reactivity of the alkyne to the extent that it can be used to protect alkynes from reduction (Eq. 8.6) or hydroboration (Eq. 8.7).[6] The alkyne can be regenerated by mild oxidative removal of the cobalt.

Eq. 8.5

$$R-C\equiv C-R + Co_2(CO)_8 \xrightarrow{\Delta}$$

Eq. 8.6

Eq. 8.7

Complexation of the alkyne not only decreases its reactivity, it also distorts its geometry towards that of an olefin. This decrease in linearity has been used to promote cyclization reactions which are sterically disfavored by the free linear alkyne (Eq. 8.8).[7]

Eq. 8.8

Complexation of alkynes to cobalt also stabilizes positive charge in the propargylic position, permitting synthetically useful reactions with propargyl cations (the Nicholas reaction) without the production of allenic byproducts (Eq. 8.9).[8] The cobalt stabilized propargyl cations can be made from a variety of precursors including propargyl alcohols (Eq. 8.9) or epoxides (Eq. 8.10)[9] by treatment with a Lewis acid. These undergo clean reaction with a variety of nucleophiles to give the propargyl substitution product, from which the cobalt is removed by mild oxidation. Cobalt complexed enynes undergo attack by acylium ions, producing the stabilized propargyl cation, which can then be trapped with nucleophiles (Eq. 8.11).[10]

Eq. 8.9

Eq. 8.10

Eq. 8.11

Although this reaction had been known for many years, it found little use in the synthesis of complex molecules until the advent of the ene-diyne antitumor agents, such as calicheamicin and neocarzinostatin.[11] The need to synthesize this unusual functional group array provided a niche for the Nicholas reaction (Eq. 8.12[11] and Eq. 8.13[12]) and many others.

Eq. 8.12

69%

Eq. 8.13

Another use of alkyne cobalt complexes is in the stereoselective aldol coupling of alkynyl aldehydes. Although free alkynyl aldehydes undergo aldol reaction with silyl enol ethers with little stereoselectivity, the complexed aldehydes react with high *syn* selectivity (Eq. 8.14).[13] Silyl ketene acetals behave similarly (Eq. 8.15)[14] and the cobalt can be readily removed by mild oxidation, giving the aldol product in good yield.

Eq. 8.14

79-86%

Eq. 8.15

R^1 = Ph, Bu, TMS

50-80%

b. The Pauson-Khand Reaction[15]

Although dicobalt complexes of alkynes are quite stable to a wide variety of electrophilic and nucleophilic reagents, when they are heated in the presence of an alkene, an interesting and useful "2+2+1" cycloaddition (called the Pauson-Khand reaction after its originators), occurs, producing cyclopentenones (Eq. 8.16).[16] The reaction joins an alkyne, an alkene, and carbon monoxide in a regioselective manner, tending to place alkene substituents next to CO and large alkyne substituents adjacent to the CO (Figure 8.1), although there are many exceptions. The mechanism of this process has not been studied, but a likely one, consistent with the products formed, is shown in Eq. 8.17 and involves the (by now) expected loss of CO to generate a vacant coordination site, coordination of the alkene, insertion of the alkene into a Co–C bond, insertion of CO, and reductive elimination.

Eq. 8.16

Eq. 8.17

Figure 8.1   Connectivity of the Pauson-Khand Reaction

Initially, the synthetic utility of the process was compromised by the necessity of severe conditions and the resulting low yields. However, it has recently been found that the reaction can be dramatically accelerated by addition of tertiary amine oxides[17] and the yields increased by immobilizing the system on polymers or on silica gel.[18] The reasons for this are obscure.

Although many intermolecular Pauson-Khand reactions have been reported, intramolecular versions are more synthetically interesting, and have been used to synthesize a variety of bicyclic systems (Eq. 8.18,[19] Eq. 8.19,[20] Eq. 8.20,[21] and Eq. 8.21[22]). With a chiral auxiliary on either the alkene or alkyne position, up to 90:10 diastereoselectivity can be achieved (Eq. 8.22).[23]

Eq. 8.18

Eq. 8.19

Eq. 8.20

Eq. 8.21

Eq. 8.22

Since stable alkyne-cobalt complexes are involved in the Pauson-Khand reaction, the ability of cobalt complexation to stabilize propargyl cations can be utilized to synthesize precursors to the Pauson-Khand reaction, greatly expanding the scope of the process. An early example utilizing thiophene as the nucleophile is shown in Eq. 8.23.[24] More recently, allyl amines (Eq. 8.24)[25] and optically active allyl boranes (Eq. 8.25)[26] have been used as nucleophiles. By taking advantage of the improved reaction conditions (R_3NO, SiO_2) mentioned earlier, good yields of complex products can be obtained. Note the high stereoselectivity of the allyl borane reaction.

Eq. 8.23

Eq. 8.24

40-90%

Eq. 8.25

50-70%
up to 88% ee

60%

Complexed enynes can act as precursors of stabilized propargyl cations, and addition of acylium ions followed by methanolysis gives precursor to polycyclic ring systems (Eq. 8.26).[27] By this procedure [5.5.5] angularly fused, [6.5.5] and [5.5.5] linearly fused, and both linear and angularly fused [6.5.5.5] and [5.5.5.5] tetracyclic compounds can be prepared.[28]

Eq. 8.26

Although intramolecular reactions tend to be quite regioselective, intermolecular reactions of disubstituted olefins often give mixtures of regioisomers. However, regiochemical control can often be regained by having a ligand — usually a heteroatom — in the homoallylic position of the alkene (Eq. 8.27).[29] A complete understanding of this process is not in hand, but substantial experimental data is available.[30]

Eq. 8.27

8.4   Metal-Catalyzed Cyclooligomerization of Alkynes

Although many transition metals catalyze the cyclotrimerization of alkynes to arenes, $CpCo(CO)_2$ is among the most efficient. The mechanism of the process has been studied extensively[31] and is shown in Eq. 8.28. It involves loss of CO from the catalyst (hence the requirement of relatively high temperatures to generate vacant coordination sites), followed by coordination of two alkynes, and formation of the metallacyclopentadiene (a "reductive" coupling of the alkynes and concomitant "oxidation" of the metal). This unsaturated metallacyclopentadiene can then either coordinate and insert another alkyne, and undergo a reductive elimination to produce the arene and regenerate the catalytically active species, or it can undergo a 4+2 cycloaddition, achieving the same result.

Initially, the process was of little synthetic interest, since it was restricted to symmetrical, internal alkynes. Terminal alkynes gave all possible regioisomers of the trisubstituted benzenes, and "crossed" cyclotrimerization of two different alkynes gave all possible products, with no control.

Eq. 8.28

An elegant solution to this latter problem and the basis of most syn-thetically useful applications of this methodology was to use a diyne as one component, and a huge alkyne, *bis*-trimethylsilyl acetylene, as the other.[32] The diyne ensures that the metallacyclopentadiene is formed only from it, since incorporation of the second alkyne unit becomes in-tramolecular. The bulk of the TMS groups prevent self-trimerization of this member. These two components cleanly cocyclotrimerize to give benzo-cyclobutanes in excellent yield. As a bonus, these are thermally cleaved to give dienes which undergo Diels-Alder reactions to form polycyclic com-pounds (Eq. 8.29). When the dienophile is built into the diyne fragment, complex products are formed expeditiously as shown by the very direct synthesis of estrone by this methodology (Eq. 8.30).[33] The protoberberine ring system is available from a longer chain diyne (Eq. 8.31).[34]

The nitrile group is another triply bonded species which does not self-trimerize, but can be cotrimerized with alkynes to give pyridines. As with simple alkyne cyclotrimerization, unsymmetric alkynes give all pos-sible regioisomers. However, cocyclotrimerization with a diyne proceeds cleanly to a single pyridine (Eq. 8.32).[35]

Eq. 8.29

Eq. 8.30

Eq. 8.31

Eq. 8.32

Alkenes will cocyclotrimerize with alkynes provided homotrimerization of the alkyne can be suppressed. This can be achieved by preforming the metallacyclopentadiene, and then adding the alkene (Eq. 8.33),[36] or by carrying the reaction out partially (Eq. 8.34)[37] or completely (Eq. 8.35[38] and Eq. 8.36[39]) intramolecularly.

Eq. 8.33

Eq. 8.34

30-60%

Eq. 8.35

Eq. 8.36

Isocyanates can fulfill the role of the double-bonded species in these cocyclotrimerizations, resulting in the production of pyridones (Eq. 8.37).[40] Nickel(0) complexes also promote the cyclocoupling of alkynes with isocyanates (Eq. 8.38).[41] This reaction proceeds through a stable metallacycle in which one alkyne and one isocyanate have been coupled. Depending on the stoichiometry another alkyne can then insert to give the pyridone, or another isocyanate can insert to give the pyrimidine. This metallacycle can also react with CO or acids to give imides or conjugated amides (Eq. 8.39).[41]

Eq. 8.37

Eq. 8.38

Eq. 8.39

Nickel(0) complexes also "reductively couple" alkynes with carbon dioxide to give stable metallacycles. These can insert another alkyne, a molecule of carbon monoxide, or undergo protolytic cleavage (Eq. 8.40).[42] This process has not yet been used to synthesize complex molecules.

Eq. 8.40

Alkynes react with many transition metals in the presence of carbon monoxide to produce compounds containing both carbon monoxide and alkyne-derived fragments. In addition, a profusion of organometallic complexes containing ligands constructed of alkyne-derived fragments and carbon monoxide are often obtained.[43] Reactions of this type are usually quite complex, and few have been controlled sufficiently to be of use in the synthesis of organic compounds. Quinones had long been known to be one of the products of the reaction of alkynes with carbon monoxide, and evidence for the intermediacy of maleoyl complexes was strong (Eq. 8.41). However these processes were of little synthetic value, because a myriad of other organic and organometallic products formed as well. By devising an efficient synthesis of maleoyl and phthaloyl complexes, and permitting these discrete complexes to react with alkynes, quinones can be made efficiently and in high yield. The most efficient phthaloyl complex is that of cobalt, prepared by the reaction of benzocyclobutanediones with CoCl(PPh₃)₃.[44] The initially formed *bis* phosphine cobalt(III) complex is unreactive toward alkynes, since a labile site axial to the phthaloyl plane is required for alkyne complexation. However, treatment with silver(I) to produce the unsaturated cationic complex[45] or better, with dimethylglyoxime in pyridine to produce an octahedral complex with a labile *cis* pyridine ligand (x-ray) leads to active phthaloyl complexes which are efficiently converted to naphthoquinones upon reaction with a variety of alkynes (Eq. 8.42). In a similar manner, benzoquinones are efficiently synthesized from cyclobutenediones (Eq. 8.43[46] and Eq. 8.44[47]). With unsymmetric systems, mixtures of regioisomers are often obtained.

Eq. 8.41

+ <u>everything</u> else
uncontrollable and messy

Eq. 8.42

Eq. 8.43

Eq. 8.44

75%

8.5 Zirconobenzyne Reactions[48]

All of the alkyne chemistry discussed above involves late transition metals, with their normal manifold of reactions. However, the early transition metals also have a rich chemistry with alkynes (see for example Chapter 4) reductively dimerizing them, among other things. Zirconium forms discrete, characterizable complexes of *benzyne*, produced by treatment of aryllithium reagents with $Cp_2Zr(Me)Cl$ followed by elimination of methane (Eq. 8.45). These reactive benzyne complexes undergo a range of insertion reactions, which ultimately result in functionalization of the arene ring. When unsymmetric aryne complexes are used, these insertions are not regiospecific but insertion into the least sterically hindered Zr–C bond is favored. Thus, isonitriles insert to give azazirconacyclopentadienes that can be cleaved to aryl ketones or iodo aryl ketones (Eq. 8.45).[49] Alkynes insert to give zirconacyclopentadienes, which are converted to benzothiophenes by treatment with sulfur dichloride.[50] Alkenes also insert to give zirconazacyclopentenes, which can be converted to a variety of substituted systems by cleavage with electrophiles.[51] Surprisingly, cyclic allylic ethers do not insert in the expected manner, but rather undergo an electrophilic cleavage to produce homoallylic alcohols (Eq. 8.46).[52]

A variety of functionalized indoles can be made from zirconabenzyne intermediates (Eq. 8.47) utilizing insertion chemistry similar to that in Eq. 8.45.[53]

Eq. 8.45

(not regiospecific X ≠ Y)
but favors insertion
adjacent to smallest group

Eq. 8.46

Eq. 8.47

References

1. (a) Reger, D.L.; Belmore, K.A.; Mintz, E.; McElligot, P.J. *Organometallics* **1984**, *3*, 134. (b) Reger, D.L.; Mintz, E. *Organometallics* **1984**, *3*, 1759.
2. Utimoto, K. *Pure Appl. Chem.* **1983**, *55*, 1845.
3. Larock, R.C.; Yum, E.K. *J. Am. Chem. Soc.* **1991**, *113*, 6689.
4. Luo, F-T.; Schreuder, I.; Wang, R-T. *J. Org. Chem.* **1992**, *57*, 2213.
5. Dickson, R.S.; Fraser, P.J. *Adv. Organomet. Chem.* **1974**, *12*, 323.
6. Nicholas, K.M..; Pettit, R. *Tetrahedron Lett.* **1971**, 3475.
7. Najdi, S.D.; Olmstead, M.M.; Schore, N.E. *J. Organomt. Chem.* **1992**, *431*, 335.
8. Padmanabhan, S.; Nicholas, K.M. *Tetrahedron Lett.* **1983**, *24*, 2239. For a review see: Nicholas K.M. *Acct. Chem. Res.* **1987**, *20*, 207.
9. Saha, M.; Nicholas, K.M. *J. Org. Chem.* **1984**, *49*, 417.
10. Smit, W.A.; Schegolev, A.A.; Gybin, A.S.; Mikaelian, G.S. *Synthesis* **1984**, 887.
11. For a review of these classes of compounds, including a discussion of the Nicholas reaction in this context see: Nicolaou, K.C.; Dai, W.M. *Angew. Chem. Int. Ed. Engl.* **1991**, *30*, 1387. Magnus, P.; Pitterna, T. *J. Chem. Soc., Chem. Comm.* **1991**, 541.
12. Magnus, P.; Fortt, S. *M. Chem. Soc., Chem. Comm.* **1991**, 544.
13. Ju, J.; Reddy, B.R.; Khan, M.; Nicholas, K.M. *J. Org. Chem.* **1989**, *54*, 5426.
14. Mukai, C.; Suzuki, K.; Nagami, K.; Hanaoka, M. *J. Chem. Soc., Perkin I* **1992**, 141.
15. For a review see: Schore, N.E. *Org. React.* **1991**, *40*. 1.
16. Billington, D.C.; Pauson, P.L. *Organometallics* **1982**, *1*, 1560.
17. Shambayati, S.; Crowe, W.E.; Schreiber, S.L. *Tetrahedron Lett.* **1990**, *31*, 5289.
18. Smit, W.D.; Kireev, S.L.; Nefedov, O.M.; Tarasov, V.A. *Tetrahedron Lett.* **1989**, *30*, 4021.
19. Exon, C.; Magnus, P. *J. Am. Chem. Soc.* **1983**, *105*, 2477.
20. (a) Billington, D.C. *Tetrahedron Lett.* **1983**, *24*, 2905. (b) Billington, D.C.; Willison, D. *Tetrahedron Lett.* **1984**, *25*, 4044.
21. Takano, S.; Inomata, K.; Ogasawara, K. *J. Chem. Soc., Chem. Comm.* **1992**, 169.
22. Takano, S.; Inomata, K.; Ogasawara, K. *Chem. Lett.* **1992**, 443.
23. Verdaguer, X.; Moyano, A.; Pericas, M.A.; Riera, A.; Greene, A.G.; Piniella, J.F.; Alvarez-Larena, A. *J. Organomet. Chem.* **1992**, *433*, 305.
24. Jaffer, H.J.; Pauson, P.J. *J. Chem. Res. (S)* **1983**, 244.
25. Jeong, N.; Yoo, S-E.; Lee, S.J.; Lee, S.H.; Chung, Y.K. *Tetrahedron Lett.* **1991**, *32*, 2137.
26. Roush, W.R.; Park, J-C. *Tetrahedron Lett.* **1991**, *32*, 6285.
27. Veretenov, A.L.; Smit, W.A.; Vorontsova, L.G.; Kurella, M.G. *Tetrahedron Lett.* **1991**, *32*, 2109.
28. Gybin, A.S.; Smit, W.A.; Caple, R.; Veretenov, A.L.; Shashkov, A.S.; Voronstova, L.G.; Kurella, M.G.; Cherthov, U.S.; Carapetyan, A.A.; Kosnikov, A.Y.; Alexanyan, M.S.; Lindeman, S.V.; Panov, V.N.; Maleev, A.V.; Struchov, V.T.; Sharpe, S.M. *J. Am. Chem. Soc.* **1992**, *114*, 5555.
29. Krafft, M.E.; Juliano, C.A.; Scott, I.L.; Wright, C.; McEachin, M.P. *J. Am. Chem. Soc.* **1991**, *113*, 1693.
30. Krafft, M.E.; Juliano, C.A. *J. Org. Chem.* **1992**, *57*, 5106.
31. (a) Wakatsuki, Y.; Kuramitsu, T.; Yamazaki, H. *Tetrahedron Lett.* **1974**, 4549. (b) McAllister, D.R.; Bercaw, J.E.; Bergman, R.G. *J. Am. Chem. Soc.* **1977**, *99*, 1666.
32. (a) Vollhardt, K.P.C. *Acc. Chem. Res.* **1977**, *10*, 1. (b) *Angew. Chem. Int. Ed. Engl.* **1984**, *23*, 539.

33. Funk, R.L.; Vollhardt, K.P.C. *J. Am. Chem. Soc.* **1980**, *102*, 5253.
34. Hillard, R.L.; Parnell, C.A.; Vollhardt, K.P.C. *Tetrahedron* **1983**, *39*, 905.
35. (a) Nalman, A.; Vollhardt, K.P.C. *Angew. Chem. Int. Ed. Engl.* **1977**, *16*, 708. (b) For a review see: Bonnemann, H.; Brijoux, W. *Adv. Het. Chem.* **1990**, *48*, 177.
36. Wakatsuki, Y.; Yamazaki, H. *J. Organomet. Chem.* **1977**, *139*, 169.
37. Grotjahn, D.B.; Vollhardt, K.P.C. *J. Am. Chem. Soc.* **1986**, *108*, 2091.
38. Butenschon, H.; Winkler, M.; Vollhardt, K.P.C. *Chem. Comm.* **1986**, 388.
39. Johnson, E.P.; Vollhardt, K.P.C. *J. Am. Chem. Soc.* **1991**, *113*, 381.
40. (a) Earl, R.A.; Vollhardt, K.P.C. *J. Org. Chem.* **1984**, *49*, 4786. (b) Earl, R.A.; Vollhardt, K.P.C. *J. Am. Chem. Soc.* **1983**, *105*, 6991.
41. (a) Hoberg, H.; Oster, B.W. *J. Organomet. Chem.* **1982**, *234*, C35. (b) Hoberg, H.; Oster, B.W. *J. Organomet. Chem.* **1983**, *252*, 359.
42. (a) Hoberg, H.; Schaefer, P.; Burkhart, G.; Kruger, C.; Ramao, M.J. *J. Organomet. Chem.* **1984**, *266*, 203. (b) Hoberg, H.; Apotecher, B. *J. Organomet. Chem.* **1984**, *270*, C15.
43. Pino, P.; Braca, G. "Carbon Monoxide Addition to Acetylenic Substrates," in *Organic Synthesis via Metal Carbonyls*; Wender, I.; Pino, P., Eds.; Wiley: New York, 1977; Vol. 2, pp 420-516; Hubel, W. "Organometallic Derivatives from Metal Carbonyls and Acetylene Compounds," in *Organic Synthesis via Metal Carbonyls*; Wender, I.; Pino, P., Eds.; Wiley: 1968; Vol. 1, pp 273-340.
44. Liebeskind, L.S.; Baysdon, S.L.; South, M.S.; Blount, J.F. *J. Organomet. Chem.* **1980**, *202*, C73.
45. Liebeskind, L.S.; Baysdon, S.L.; South, M.S.; Iyer, S.; Leeds, J.P. *Tetrahedron* **1985**, *41*, 5839.
46. Liebeskind, L.S.; Jewell, C.F. *J. Organomet. Chem.* **1985**, *285*, 305.
47. Liebeskind, L.S.; Chidambaram, R.; Nimkar, S.; Liotta, D. *Tetrahedron Lett.* **1990**, *31*, 3723.
48. For a review see: Buchwald, S.L.; Nielsen, R.B. *Chem. Rev.* **1988**, *88*, 1047.
49. Buchwald, S.L.; King, S.M. *J. Am. Chem. Soc.* **1991**, *113*, 258.
50. Buchwald, S.L.; Fang, Q. *J. Org. Chem.* **1989**, *54*, 2793.
51. Cuny, G.D.; Gutierrez, A.; Buchwald, S.L. *Organometallics* **1991**, *10*, 537.
52. Cuny, G.D.; Buchwald, S.L. *Organometallics* **1991**, *10*, 363.
53. Tidwell, J.H.; Senn, D.R.; Buchwald, S.L. *J. Am. Chem. Soc.* **1991**, *113*, 4685.

CHAPTER 9

Synthetic Applications of η³-Allyl Transition Metal Complexes

9.1 Introduction

Although η³-allyl complexes are known for virtually all of the transition metals, relatively few have found use in organic synthesis. However, those that have, mainly those of Pd, have broad utility. η³-Allyl metal complexes can be made from a wide range of organic substrates in a variety of ways (Figure 9.1). These include (1) oxidative addition of allylic substrates to metal(0) complexes, (2) insertion of 1,3-

Figure 9.1 Preparation of η³-allylmetal Complexes

dienes into a metal hydride (or metal alkyl), (3) allylic proton abstraction from a π-olefin complex, (4) nucleophilic attack on a 1,3-diene metal complex, and (5) reaction of Grignard (or other main group metal) allyl compound with transition metal halides. Although many η^3-allyl complexes are stable and isolable, they also are quite reactive, under appropriate conditions. The scope of reactions in which these complexes participate is presented in the following sections.

9.2 Transition-Metal Catalyzed Telomerization of 1,3-Dienes

Treatment of 1,3-dienes with nucleophiles in the presence of palladium(0) catalysts results in the production of functionalized octadienes, made by the joining of two 1,3-dienes with incorporation of the nucleophile (Eq. 9.1).[1] The mechanism of this process has not been closely studied but is thought to involve complexation of two dienes to the Pd(0) followed by the (by now) familiar "reductive dimerization" of the diene units (actually an oxidative addition of two dienes to the metal) joining them and generating a *bis*-η^3-allylpalladium species. As we shall soon see, η^3-allylpalladium complexes are generally subject to nucleophilic attack, usually at the less substituted terminus. Nucleophilic attack produces an anionic π-olefin-η^3-η^1 allyl complex which undergoes proton transfer to the metal followed by reductive elimination to produce the diene "telomer" (an imprecise term used here to denote the combination of two diene units with a nucleophile) (Eq. 9.2). This procedure offers a convenient way to assemble functionalized chains (Eq. 9.3),[2] but it has found only modest use in complex molecule synthesis.

Eq. 9.1

Eq. 9.2

Start here

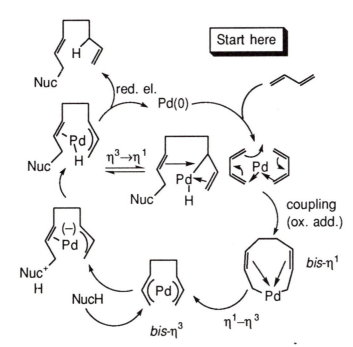

NucH = AcOH, H$_2$O, ROH, RNH$_2$ $\overset{Y}{\underset{X}{<}}$ where X, Y = CO$_2$Et, CN, NO$_2$, COR

Eq. 9.3

The synthetic utility of the above process was dramatically enhanced when it was applied in an intramolecular fashion (Eq. 9.4),[3] assembling five-membered rings with appended functionality quickly and efficiently. Provided the substrate has appropriately situated β-hydrogens, this cyclization can take place without nucleophilic attack, producing trienes instead of dienes (Eq. 9.5).[4] This process is closely related to the eneyne cyclizations proceeding by "hydrometallation" (Chapter 3), as well as to a number of other cyclization reactions catalyzed by low valent transition metals.

Eq. 9.4

NucH = PhCH$_2$OH, PhOH, Et$_2$NH, pTsOH, CH$_3$NO$_2$, CH$_2$(CO$_2$Et)$_2$, R$_3$SiH,

Eq. 9.5

Most closely related are the iron(0) coupling reactions as exemplified by Eq. 9.6[5], Eq. 9.7[5], and Eq. 9.8[6]. Again, the mechanism has not been studied but is likely to be similar to that in Eq. 9.5 (e.g., Eq. 9.9).

Eq. 9.6

85%

Eq. 9.7

Eq. 9.8

Eq. 9.9

Perhaps the earliest example of this type of process is the commercially important nickel(0) catalyzed cyclooligomerization of 1,3-dienes to produce a wide variety of compounds depending on conditions (Figure 9.2).[7]

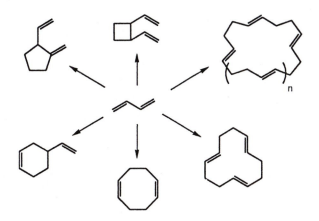

Figure 9.2   Ni(0) Catalyzed Cyclooligomerization of Butadiene

Whatever the ultimate product, they all derive from an initial dimerization of the diene to form the *bis*-η^3-allyl nickel complex (much as in Eq. 9.2 with Pd). Insertion of an additional diene leads to the cyclic trimers, while reductive elimination from one of the several bis-η^1-allyl species produces cyclooctadienes, vinylcyclohexanes or divinylcyclobutanes (Eq. 9.10). Although this chemistry is very well developed with simple sub- strates, its real synthetic utility for complex systems resides in in- tramolecular reactions directed toward forming eight-membered rings, such as in the synthesis of (+)-astericanolide (Eq. 9.11).[8] Alkynes are also effi- cient partners in a similar process for forming bicyclic compounds (Eq. 9.12).[9]

Eq. 9.10

Eq. 9.11

Eq. 9.12

9.3 Palladium Catalyzed Reactions of Allylic Substrates

a. General[10]

Palladium(0) complexes catalyze a wide variety of synthetically useful reactions of allylic substrates which proceed (Figure 9.3) through η^3-allyl intermediates. Whatever the ultimate product, they all involve the same manifold of familiar steps, and most involve nucleophilic attack on the η^3-allyl palladium complex. The reaction begins by the oxidative addition of the allylic substrate to the palladium(0) catalyst (a), a process that goes with clean inversion to initially give the η^1-allyl complex which is in equilibrium with the η^3-allyl complex (b) and is rarely detected. In contrast, the η^3-allyl complexes are quite stable, yellow solids which can easily be isolated if desired. However in the presence of excess ligand, low equilibrium concentrations of a cationic η^3-allyl complex (c) are generated. This is very reactive towards a wide range of nucleophiles, and undergoes nucleophilic attack from the face opposite the metal (d) (inversion) to produce the allylated nucleophile, and to regenerate palladium(0) (e) to reenter the catalytic cycle. The net stereochemistry of this allylic substitution is retention, which is the result of two inversions. The regiochemistry with unsymmetric allyl substrates, in general, favors attack at the less substituted allyl terminus, but is somewhat dependent on the nucleophile and the metal, as well as the ligands.

η^3-Allyl complexes undergo transmetallation with main group organometallics generating an η^3-allyl-η^1-alkyl palladium(II) complex (f), which undergoes reductive elimination (g) to produce the alkylated allyl compound. Since, in this case, the nucleophile (R' of R'M) is delivered first to the metal and then to the η^3-allyl group from the *same* face as the metal, this step occurs with retention, and the overall alkylation goes with net inversion. η^3-Allyl complexes can also insert alkenes (h) (probably via their η^1-isomers), producing 1,5-dienes after β-hydrogen elimination. All of these processes have been applied to the synthesis of complex molecules and are discussed in detail below.

A number of palladium(0) complexes and palladium(II) precatalysts can be used, and the choice of catalyst is often not important. The most frequently utilized catalyst is (Ph$_3$P)$_4$Pd, which is commercially available. However, commercial material is often of widely varying activity and it is best to prepare one's own. η^3-Allyl palladium(II) complexes are themselves catalysts, and because of their ease in preparation and handling, they are often used. A very convenient catalyst results from the treatment of the air-stable, easily prepared and handled Pd(dba)$_2$ (dba = PhCH=CH–CO–CH=CHPh, dibenzylidene acetone) with varying amounts of triphenylphosphine, generating PdL$_n$ *in situ*. Finally, a variety of palladium(II) salts are readily reduced in the presence of substrate and ligand, and are often efficient precatalysts for the process.

Figure 9.3   Palladium Catalyzed Reactions of Allylic Substrates

b. Allylic Alkylation[11]

Alkylation of allylic substrates by stabilized carbanions is one of the synthetically most useful reactions catalyzed by palladium(0) complexes (Eq. 9.13). A wide range of allylic substrates undergo this reaction with a reasonable range of carbanions, making this an important process for the formation of carbon-carbon bonds. The reaction is very stereoselective, and proceeds with overall retention of configuration, the result of two inversions (Eq. 9.14).[12] The reaction is also quite regioselective, with attack at the less substituted terminus of the η^3-allyl intermediate favored, regardless of the initial position of the allylic leaving group.

Eq. 9.13

$Z = Br, Cl, OAc, -OCOR, OP(OEt)_2, O-S-R, OPh, OH, R_3N^+, NO_2, SO_2Ph, CN,$

$X, Y = CO_2R, COR, SO_2Ph, CN, NO_2$

Eq. 9.14

(both catalytic and stoichiometric processes go with clean retention)

Because of the difference in reactivity of various allylic functional groups (e.g., Cl > OCO$_2$R > OAc >> OH) high regioselectivity can be achieved with *bis* allyl substrates, in addition to high stereoselectivity (Eq. 9.15).[13] Asymmetric induction using optically active ligands has proved difficult, although a number of specific systems have been developed (Eq. 9.16[14] and Eq. 9.17[15]). One complication to asymmetric induction is the recent observation that optically active η^3-allyl palladium complexes are rapidly racemized by excess Pd(0) catalyst (by Pd–Pd exchange with inversion each time).[16] The real utility of this process lies in the range of functionality and complexity tolerated in both participants in the reaction, as exemplified in Eq. 9.18,[17] Eq. 9.19,[18] and Eq. 9.20.[19]

Eq. 9.15

Eq. 9.16

Eq. 9.17

up to 82% ee

Eq. 9.18

Eq. 9.19

83%

Eq. 9.20

Control of stereochemistry in acyclic systems is among the most difficult of problems faced in organic synthesis. The stereospecific nature of both the formation of η^3-allyl palladium complexes and their reactions with carbanions has been used to relay the stereochemistry of one chiral center to a remote position in conformationally mobile systems.[20] Thus, organopalladium chemistry is used to transfer the chirality of a vinyl lactone moiety to the remote vinylic carbon, producing a product having two chiral centers in a 1,5 relationship (Eq. 9.21). For this process to succeed, formation of the η^3-allyl palladium complex must proceed only from the conformation shown, the η^3-allyl complex must maintain its stereochemistry, and the nucleophile must attack regiospecifically at the terminal carbon of the η^3-allyl system. All of these criteria are met, and the reaction goes in 90% yield with greater than 95% stereoselectivity.

Eq. 9.21

Intramolecular versions of this process are very efficient for the formation of small (Eq. 9.22),[21] normal (Eq. 9.23,[22] Eq. 9.24,[23] and Eq. 9.25[24]), medium (Eq. 9.26[25] and Eq. 9.27[26]), and large rings (Eq. 9.28).[27] Note that for medium rings (Eq. 9.27) the ring size can be altered by using chelating ligands, although this is not a general phenomenon.

Eq. 9.22

Eq. 9.23

Eq. 9.24

80%

Eq. 9.25

77%

with L = (R,S) BPPFA, 34% yield, 83% ee

Eq. 9.26

49%

Eq. 9.27

dppe = Ph₂P⌒⌒⌒PPh₂

L = Ph₃P   38%   93:7
L = dppe   51%   5:95

Eq. 9.28

92%

c. Allylic Amination, Alkoxylation, Reduction, and Deprotonation

The reactions of nitrogen nucleophiles with allylic substrates are among the most useful for organic synthesis, and both 1° and 2° amines (but *not* NH$_3$) efficiently aminate allylic substrates, as do sulfonamide anions. The stereo- and regioselectivity of allylic amination parallels that of allylic alkylation, and the range of reactive substrates is comparable. The nitrogen almost invariably ends up at the less hindered terminus of the allyl system, although that may not be the initial site of attack, since Pd(0) complexes catalyze the allylic transposition of allyl amines.

Intermolecular aminations are efficient and a range of functionality is tolerated (Eq. 9.29[28] and Eq. 9.30[29]). Current interest in aminoglycosides (Eq. 9.31)[30] and nucleoside analogs (Eq. 9.32)[31] has resulted in the application of palladium(0) catalyzed allylic amination to these problems. Azide is readily introduced into allylic epoxides under similar conditions (Eq. 9.33).[32]

Eq. 9.29

Eq. 9.30

Eq. 9.31

36-80%

$R_2NH =$

Eq. 9.32

54%

Eq. 9.33

aq THF

Pd(dba)$_2$/2 PPh$_3$
25 °C, 10 min

75-85%

Intramolecular aminations are particularly effective, and a wide range of fused (Eq. 9.34),[33] spiro (Eq. 9.35),[34] and macrocyclic systems (Eq. 9.36)[35] can be made efficiently.

Eq. 9.34

1.5% Pd$_2$(dba)$_3$·CHCl$_3$

12% P(OR)$_3$
H$_2$O/THF
65°C

73%

Eq. 9.35

5% Pd(PPh₃)₄

Et₃N/CH₃CN
70 °C, 2 h

>90%

Eq. 9.36

10% Pd(PPh₃)₄
dppb, THF
70 °C high
dilution

89%

The use of oxygen nucleophiles in these allylic reactions is considerably less common, although those few that are reported (Eq. 9.37[36] and Eq. 9.38[37]) are efficient. One of the most synthetically useful systems involving oxygen nucleophiles is the palladium catalyzed *bis* acetoxylation or chloroacetoxylation of 1,3-dienes (Eq. 9.39)[38] briefly introduced in Chapter 7. With palladium acetate as catalyst, the initial acetoxylation of the olefin takes place with *trans* stereochemistry, consistent with external nucleophilic attack on an olefin Pd(II) complex. In the absence of added chloride, oxidative cleavage of the η³-allyl palladium complex results in a *cis* migration of coordinated acetate to the η³-allyl ligand, giving the *trans* diacetate and regenerating the palladium(II) acetate catalyst. In contrast, in the presence of catalytic amounts of added chloride, oxidative cleavage results in attack of uncoordinated acetate from the face opposite the palladium, giving the *cis* diacetate cleanly.[39] This is a general process and more than twenty cyclic and acyclic dienes have been studied. By using larger amounts of chloride, the above reaction can be converted to an efficient synthesis of 1-acetoxy-4-chloro-2-alkenes. The reaction is stereoselective in this case as well, and proceeds by a *trans* acetoxypalladation followed by an oxidative displacement of palladium by chloride from the face opposite the metal. This is a very useful process, since it allows the chemical differentiation of the two termini of the diene, and further, gives a substrate with two positions potentially accessible to η^3-allyl palladium chemistry. (Allylic halides are more reactive than allylic acetates.[40])

Eq. 9.37

Eq. 9.38

Eq. 9.38

This chemistry has been used to develop quite efficient and stereoselective approaches to fused tetrahydrofurans and tetrahydropyrans (Eq. 9.40).[41] In this case, the tethered alcohol group is the nucleophile which initially attacks the π-olefin complex, giving the *trans* η³-allyl complex. Inter- or intramolecular attack of acetate, chloride, or alkoxide on this η³-allyl complex leads to the observed products. Other applications of this very useful process will be presented later in this chapter.

Eq. 9.40

n = 1,2   m = 1,2   60-90%

Palladium(0) complexes catalyze a number of other useful reactions of allylic substrates. Allyl acetates are easily reduced to alkenes, usually with allylic transposition, in the presence of a hydride source which attacks the η³-allyl palladium intermediate. External reducing agents such as polymethylhydrosiloxane (Eq. 9.41),[42] cyanoborohydride or triethyl borohydride all work well. Allyl formates undergo intramolecular reduction, with formate acting as the hydride source (Eq. 9.42).[43] In these cases the hydride is delivered from the same face as the palladium. In the absence of nucleophiles, palladium(0) will catalyze elimination reactions, converting allylic substrates to dienes (Eq. 9.43[44] and Eq. 9.44[45]).

Eq. 9.41

Eq. 9.42

n = 1,2

and

Eq. 9.43

78-91%
8 cases

Eq. 9.44

93%

and

96%

Finally, allylic carbonates and esters make excellent protecting groups for carboxylic acids, alcohols and amines, because they are stable to a wide range of reaction conditions, and can be exclusively removed in the presence of other labile protecting groups (Eq. 9.45,[46] Eq. 9.46,[47] and Eq. 9.47[48]). In these cases, a nucleophile must be present to regenerate the catalyst from the initially formed η^3-allyl complex.

Eq. 9.45

Eq. 9.46

Eq. 9.47

d. Allylic Alkylation by Transmetallation

Transmetallation has been used much less frequently in allylic systems than with aryl or vinyl systems, for several practical reasons. Allyl acetates are the most attractive class of allylic substrates, because of their ready availability from the corresponding alcohols, but they are substantially less reactive towards oxidative addition to Pd(0) than are allylic halides, and phosphine ligands are required to promote this process. Acetate coordinates quite strongly to palladium and this, along with the presence of excess phosphine, slows down the crucial transmetallation step, already the rate limiting step. Finally, in contrast to dialkyl palladium(II) complexes, σ-alkyl-π-allyl palladium(II) complexes undergo reductive elimination only slowly, compromising this final step in the catalytic cycle. It is thus not surprising that alkylation of allylic acetates via transmetallation has been slow to develop. However under appropriate reaction conditions (polar solvents such as DMF, "ligandless" (no phosphine) catalysts such as Pd(dba)$_2$ or PdCl$_2$(MeCN)$_2$, excess LiCl to facilitate transmetallation) allyl acetates can be alkylated by a variety of aryl and vinyl tin reagents (e.g., Eq. 9.48).[49] Coupling occurs at the less substituted terminus of the allyl system with clean inversion (inversion in the oxidative addition step, retention in the transmetallation/reductive elimination step), and the geometry of the alkene in both the allylic substrate and vinyl tin partner is maintained.

Eq. 9.48

Allyl epoxides are better substrates for this process, and are alkylated by a variety of aryl and vinyl tin reagents. Acetylenic, benzylic or allylic tin compounds fail to couple (Eq. 9.49[50] and Eq. 9.50[51]). Allylic chlorides are the most reactive of substrates, undergoing this coupling reaction under "normal" reaction conditions (phosphine present) in excellent yield with highly functionalized substrates (Eq. 9.51[52] and Eq. 9.52[53]). Intramolecular versions have also been developed (Eq. 9.53).[54]

Eq. 9.49

Eq. 9.50

e. Insertion Processes Involving
η^3-Allyl Palladium Complexes

Alkenes, alkynes, and carbon monoxide insert into η^3-allyl palladium complexes (perhaps via the η^1 isomer), generating σ-alkyl palladium complexes which can undergo the very rich chemistry described in Chapter 4. When carried out in an intramolecular manner efficient cyclization processes can be developed (Eq. 9.54).[55]

Eq. 9.54

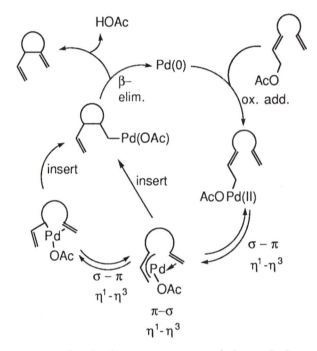

These reactions are synthetically very powerful, and demonstrate the real potential for the use of transition metals in organic synthesis, since often even the substrates are synthesized using transition metals. For example, in Eq. 9.55, the 1-acetoxy-4-chlorobut-2-ene was synthesized by the Pd(OAc)$_2$ catalyzed chloroacetoxylation of dienes (Eq. 9.39) and the diallyl *bis* sulfone prepared by Pd-catalyzed allylic alkylation of allyl halides. Treatment of this compound with Pd(dba)$_2$ in acetic acid results in efficient cyclization. The reaction is quite general, and tolerates a variety of functional groups (Eq. 9.56–Eq. 9.59). When additional unsaturation is appropriately situated, additional insertions can occur, resulting in the formation of several rings in a single efficient process (Eq. 9.60[56] and Eq. 9.61[57]).

Eq. 9.55

Eq. 9.56

Eq. 9.57

Eq. 9.58

Eq. 9.59

Eq. 9.60

Eq. 9.61

80%

Dienes also insert into η^3-allyl palladium complexes, producing another η^3-allyl complex, which can undergo nucleophilic attack by acetate to form diene acetates (Eq. 9.62 and Eq. 9.63).[58] Again, palladium catalysis is also used to synthesize the starting materials. In a related process, bicyclic amines were synthesized by insertion of an alkene into an η^1-vinyl palladium complex, producing an η^3-allyl complex which was trapped by a pendent tosamide (Eq. 9.64).[59]

Eq. 9.62

Eq. 9.63

56%

Eq. 9.64

Insertion of an alkene into an η^3-allyl palladium complex generates an η^1-alkyl palladium complex, which itself is very reactive towards insertion of carbon monoxide (Chapter 4). By carrying out these cyclizations under an atmosphere of carbon monoxide the η^1-alkyl palladium intermediate can be intercepted making an η^1-acyl palladium complex. Trapping with methanol results in formation of esters (Eq. 9.65).[60] If methanol is left out of the reaction mixture the η^1-acyl complex can insert the alkene from the original allyl acetate to form yet another ring (Eq. 9.66).[61] With appropriate adjustment of substrate structure and reaction conditions, impressive polycyclizations can be achieved (Eq. 9.67[61,62] and Eq. 9.68[63]).

Eq. 9.65

Eq. 9.66

Eq. 9.67

Eq. 9.68

f. Palladium(0) Catalyzed Cycloaddition Reactions via Trimethylenemethane Intermediates

Although stable trimethylenemethane complexes of transition metals have been known for a long time, they have found little use in synthesis because of their lack of reactivity. However, 1,1'-bifunctional allyl compounds having an allylic acetate and an allyl silane undergo reaction with palladium(0) complexes to produce unstable, uncharacterized intermediates that react as if they were zwitterionic trimethylenemethane complexes, undergoing 3+2 cycloadditions to a range of electron deficient alkenes (Eq. 9.69).[64] The reaction is thought to involve oxidative addition of the allyl acetate to the Pd(0) complex, generating a cationic η^3-allyl palladium complex. Displacement of the remaining allyl silane group, perhaps by acetate, produces a zwitterionic trimethylene methane complex, the anionic end of which attacks the β-position of a conjugated enone, generating an α-enolate which in turn attacks a terminus of the electrophilic cationic η^3-allyl palladium complex.

Eq. 9.69

It is not known if this cycloaddition is concerted; however, since the stereochemistry of the alkene partners is maintained in the product, ring closure must occur faster than rotation about an incipient single bond. In addition, the process is quite diastereoselective (Eq. 9.70 and Eq. 9.71).[65]

Eq. 9.70

Eq. 9.71

With unsymmetrical bifunctional allyl acetates, regioselective coupling occurs, with coupling at the more substituted terminus regardless of the electronic nature of the functional group and the initial position of the acetate and silyl leaving groups (Eq. 9.72 and Eq. 9.73).[66] This indicates that equilibration of the three termini of the trimethylmethane fragment must occur more rapidly than does coupling. With more extended acceptors such as pyrones or tropone, 3+2, 3+4, or 3+6 cycloadditions may occur, depending on the substituents on the acceptor (Eq. 9.74 and Eq. 9.75).[67] Intramolecular versions of this reaction are quite efficient for the production of highly functionalized polycyclic compounds (Eq. 9.76,[68] Eq. 9.77,[69] and Eq. 9.78 and Eq. 9.79[70]).

Eq. 9.72

Eq. 9.73

Eq. 9.74

but

Eq. 9.75

Eq. 9.76

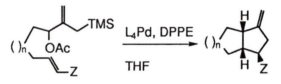

n = 1, 2, 6
Z = CO₂Me, SO₂Ph

Eq. 9.77

n = 2, 3, 4, 8

Eq. 9.78

Eq. 9.79

100%

Aldehydes can also participate in [3+2] cycloaddition reactions with bifunctional allyl acetates in the presence of palladium(0) catalyst, forming five-membered oxygen heterocycles (Eq. 9.80).[71] With unsymmetrical trimethylene methane precursors, mixtures of regioisomers are obtained, in contrast to the observations with conjugated enones. This is rationalized by asserting that cyclization competes effectively with rearrangement of the initially formed trimethylene methane complexes. Tin cocatalysts dramatically improve the yields and regioselectivity of the process, perhaps by acting as a Lewis acid to increase the reactivity of the aldehyde. A more profound effect is noted when an In(III) cocatalyst is used with conjugated enones as acceptors. This additive completely changes the regioselectivity of this reaction, from 1,4 (addition to the olefin) to 1,2 (addition to the carbonyl group) (Eq. 9.81).[72] This change is thought to be the result of coordination of the highly electropositive In(III) to the carbonyl oxygen, enhancing reaction at this site over attack of an olefinic position.

Eq. 9.80

R = Me, Ph, CH=CH₂, CN, CEt, OAc

Eq. 9.81

80%   PPh₃   In(acac)₃   "1,2"   70%

Figure 9.3 "Trimethylene Methane" Reactions of Methylene Cyclopropenes

Methylenecyclopropanes also undergo palladium(0) catalyzed [3+2] cycloaddition reactions, again, most likely through trimethylene methane intermediates (Figure 9.3)[73] formed by "oxidative addition" into the activated cyclopropane carbon–carbon bond. Nickel(0) complexes catalyze similar processes. Although most of the studies of this system have been carried out on relatively simple substrates, a few, more complex, systems have been examined (Eq. 9.82).[74]

Eq. 9.82

9.4 η^3-Allyl Complexes of Metals Other than Palladium

a. Molybdenum and Tungsten

Molybdenum and tungsten hexacarbonyl both catalyze the alkylation of allylic acetates by stabilized carbanions. However, the regioselectivity is quite different from that observed with palladium(0) (Figure 9.4).[75] Whereas palladium directs the nucleophile to the less substituted terminus of the η^3-allyl system, molybdenum leads to reaction at the more substituted end when malonate is the attacking anion, and the less substituted end when more sterically demanding nucleophiles are used. Tungsten catalysts result in attack at the more substituted terminus, regardless of the steric bulk of the nucleophile. The complex $Mo(RNC)_4(CO)_2$ is a more efficient catalyst than $Mo(CO)_6$, but, in this case, attack at the less substituted allyl terminus predominates.[76]

Preformed η^3-allylmolybdenum complexes can be used both to activate adjacent functional groups and to control stereochemistry. For example, the acetyl complex in Eq. 9.83 undergoes facile aldol condensation with benzaldehyde followed by reduction by $NaBH_4$ from the face opposite the metal. Replacement of the relatively inert CO ligands by nitrosyl (formally +) and chloride produces a more labile complex which slowly condenses with benzaldehyde to give the 1,3-diol in modest yield after hydrolysis (Eq. 9.83).[77]

Figure 9.4   Group 6 Metal-Catalyzed Reactions of Allyl Acetates

Eq. 9.83

b. η³-Allyliron Complexes

Cationic η³-allyliron tricarbonyl undergo nucleophilic attack by stabilized carbanions in much the same way as η³-allylpalladium complexes, and NaFe(CO)₃(NO) will catalyze the allylic alkylation of allyl chlorides and acetates.[78] However, perhaps because of the efficiency of the Pd-catalyzed processes, this related iron chemistry has found little use in organic synthesis. Preformed η³-allyliron complexes are alkylated by main group η¹-allyl metals (Eq. 9.84)[79] and by silylenol ethers (Eq. 9.85).[80] The synthetic potential of these reactions has yet to be established.

Eq. 9.84

Eq. 9.85

c. η^3-Allylcobalt Complexes

The cobalt carbonyl anion $[Co(CO)_4]^-$ is a weak base but modest nucleophile which reacts with allylic halides to produce η^3-allylcobalt carbonyl complexes (Eq. 9.86).[81] These are relatively unstable, deep red oils, and little synthetic use for them has been found. When treated with methyl iodide, the $[Co(CO)_4]^-$ anion reacts to form the σ-methyl complex, which readily inserts CO to form the η^1-acyl complex. Treatment of this complex with butadiene results in another insertion, producing the β-acyl-η^3-allyl cobalt complex (Eq. 9.87). Treatment with a strong base abstracts the quite acidic α-proton, producing the acyl diene, and regenerating the cobalt carbonyl anion.[82] In contrast, treatment with a stabilized carbanion results in nucleophilic attack at the unsubstituted end of the η^3-allyl group, resulting in an overall alkylation-acylation of the 1,3-diene.[83] Neither of these processes have been utilized with complex substrates.

Eq. 9.86

Eq. 9.87

d. η³-Allylnickel Complexes[84]

η³-Allylnickel halides are generated in high yield by the reaction of allylic halides with nickel(0) complexes, usually nickel carbonyl or *bis*(cyclooctadiene)nickel in nonpolar solvents such as benzene (Eq. 9.88). This reaction tolerates a range of functional groups in the allyl chain, and allows the preparation of a number of synthetically useful functionalized complexes. η³-Allylnickel halides are not directly accessible from olefins, in contrast to the corresponding palladium complexes, at least in part because nickel(II)-olefin complexes are virtually unknown, allowing no pathway for activation of the allylic position for proton removal. η³-Allylnickel halide complexes are deep red to red-brown crystalline solids, which are quite air-sensitive in solution, but are stable in the absence of air.

Although η³-allylpalladium halides are subject to nucleophilic attack, η³-allylnickel halides are not. Instead they behave, at least superficially, as if they were nucleophiles themselves, reacting with organic halides, aldehydes, and ketones, to transfer the allyl group. However, these reactions are radical-chain processes rather than nucleophilic reactions, and the chemistry of η³-allylnickel halides is drastically different from that of the corresponding palladium complexes.

Eq. 9.88

red, air-sensitive
solid

The best-established and most widely used reaction of η³-allyl-nickel halide complexes is their reaction with organic halides to replace the halogen with the allyl group (Eq. 9.89). This reaction proceeds only in polar coordinating solvents, such as DMF, HMPA, or N-methylpyrroli-done.

Eq. 9.89

R = alkyl, aryl, vinyl, benzyl

tolerates OH, NH$_2$, CO$_2$R, CO$_2$H, CHO, COR, CN;
I > Br > Cl

Aryl, vinyl, primary, and secondary alkyl bromides and iodides react in high yields, with aryl and vinyl halides being considerably more reactive than the alkyl halides. Chlorides react much more slowly than bromides or iodides. This reaction tolerates a wide variety of functional groups including hydroxyl, ester, amide, and nitrile. These complexes will react with bromides in preference to chlorides in the same molecule, and will tolerate ketones and aldehydes in some instances. With unsymmetric η^3-allyl groups, coupling occurs exclusively at the less-substituted terminus, in contrast to most main group allyl organometallics. This property was used to an advantage in the synthesis of α-santalene, and epi-β-santalene (Eq. 9.90).[85] Although allylic halides are among the most reactive toward these complexes, coupling normally results to give all possible products because of rapid exchange of η^3-allyl ligand with the allyl halide. If the two allyl groups are somewhat different electronically, selective cross-coupling is sometimes observed in reasonable yield.[86] The

Eq. 9.90

stereochemistry of the olefin in the η^3-allyl system is lost in these allyl transfer reactions, but the stereochemistry of the olefin in vinyl halide substrates is normally maintained.

The ability of η^3-allylnickel halides to react with aryl halides under very mild conditions and to tolerate a wide range of functional groups permits the introduction of allyl groups into a very broad array of substrates (Eq. 9.91,[87] Eq. 9.92,[86a] Eq. 9.93,[88] and Eq. 9.94[89]).

Eq. 9.91

Eq. 9.92

Eq. 9.93

Eq. 9.94

Conjugated enones react with nickel(0) complexes in the presence of trimethylsilyl chloride to produce 1-trimethylsilyloxy-η^3-allylnickel complexes. These undergo typical coupling reactions with organic halides to produce silylenol ethers (Eq. 9.95).[90] This product is one that would arise by nucleophilic (R-) addition to the β-position of the starting enone.

Eq. 9.95

R^1 = H, Me, Ph
R^2 = H, Me

80-97%

60-80%

R = nPr, nBu, ..., Ph, iPr, pCHOPh

However, by complexation to nickel, the normal reactivity patterns are reversed, and the β-alkyl group is introduced as an electrophile (RX = "R⁺"). By using η^3-allylnickel complexes as intermediates, the conjugate addition of alkyltin reagents to conjugated enones can be catalyzed. Although these two processes are related, they differ somewhat mechanistically (Eq. 9.96).[91]

Eq. 9.96

48-79%

$R = $ ..., Ph ..., ..., $CH_3-\overset{O}{\underset{\|}{C}}-$

Although the products of the reaction between η^3-allylnickel halides and organic halides are simple, the process by which they form is complex. The process appears to be a radical-chain reaction, initiated by light or added reducing agents and completely inhibited by less than one mole percent of *m*-dinitrobenzene, an efficient radical anion scavenger. Chiral secondary halides racemize upon allylation, implying the intermediacy of free carbon-centered radicals, but vinyl halides maintain their streochemistry, implying an absence of free carbon-centered radicals.[92]

Although η^3-allylnickel halides react with organic halides in preference to aldehydes or ketones, under slightly more vigorous conditions (50°C versus 20°C) carbonyl compounds do react to produce homoallylic alcohols.[93] α-Diketones are the most reactive substrates, producing α-ketohomoallylic alcohols. Aldehydes and cyclic ketones, including cholestanone, progesterone, and 5-α-androstane-3,17-dione, react well (the steroids at the most reactive carbonyl group), but simple aliphatic and α,β-unsaturated ketones react only sluggishly. Again, reaction occurs at the less-substituted end of the allyl group, in contrast to main group allyl organometallics. The reaction of η^3-(2-carboethoxyallyl)nickel bromide with aldehydes and ketones produces α-methylene-γ-butyrolactones (Eq. 9.97).

Eq. 9.97

References

1. (a) Tsuji, J. *Acct. Chem. Res.* **1973**, *6*, 8. (b) Tsuji, J. *Pure Appl. Chem.* **1981**, *53*, 2371. (c) Tsuji, J. *Pure Appl. Chem.* **1982**, *54*, 197.
2. Takahashi, T.; Minami, I.; Tsuji, J. *Tetrahedron Lett.* **1981**, *22*, 2651.
3. (a) Takacs, J.M.; Zhu, J. *J. Org. Chem.* **1989**, *54*, 5193. (b) Takacs, J.M.; Chandramouli, S. *Organometallics* **1990**, *9*, 2877. (c) Takacs, J.M.; Zhu, J. *Tetrahedron Lett.* **1990**, *31*, 1117.
4. Takacs, J.M.; Zhu, J.; Chandramouli, S. *J. Am. Chem. Soc.* **1992**, *114*, 773.
5. Takacs, B.E.; Takacs, J.M. *Tetrahedron Lett.* **1990**, *31*, 2865.
6. Takacs, J.M.; Myoung, Y.C.M. *Tetrahedron Lett.* **1992**, *33*, 317.
7. Jolly, P.W. "Nickel-Catalyzed Oligomerization of Alkenes and Related Reactions" in *Comprehensive Organometallic Chemistry*, Wilkinson, G.; Stone, F.G.A.; Abel, E.W. eds., Pergamon, New York, 1982, vol. 6, pp. 615-48.
8. Wender, P.A.; Ihle, N.C.; Corriea, C.R.D. *J. Am. Chem. Soc.* **1988**, *110*, 5904.
9. Wender, P.A.; Jenkins, T.E. *J. Am. Chem. Soc.* **1989**, *118*, 6432.
10. For a recent review on selectivity in these reactions see: Frost, C.G.; Howarth, J.; Williams, J.M.J. *Tetrahedron Asymm.* **1992**, *3*, 1089.
11. For reviews see: (a) Trost, B.M.; Verhoeven, T.R. *J. Am. Chem. Soc.* **1980**, *102*, 4730. (b) Yamamoto, T.; Saito, O.; Yamamoto, A. *J. Am. Chem. Soc.* **1981**, *103*, 5600. (c) Trost, B.M. *Acc. Chem. Res.* **1980**, *13*, 385.
12. (a) Hayashi, T.; Konishi, M.; Kumada, M. *J. Chem. Soc., Chem. Comm.* **1984**, 107. (b) Hayashi, T.; Hagihara, T.; Konishi, M.; Kumada, M. *J. Am. Chem. Soc.* **1983**, *105*, 7767. (c) Leutenegger, V.; Umbricht, G.; Fahrni, C. von Matt, P.; Pfaltz, A. *Tetrahedron* **1992**, *48*, 2143.
13. Schink, H.E.; Backvall, J-E. *J. Org. Chem.* **1992**, *57*, 1588.
14. Sawamura, M.; Nagata, H.; Sakamoto, H.; Ito, Y. *J. Am. Chem. Soc.* **1992**, *114*, 2586.
15. Okada, Y.; Minami, T.; Sasaki, Y.; Umezu, Y.; Yamaguchi, M. *Tetrahedron Lett.* **1990**, *31*, 3905.
16. Granberg, K.L.; Backvall, J-E. *J. Am. Chem. Soc.* **1992**, *114*, 6858.
17. Eren, D.; Keinan, E. *J. Am. Chem. Soc.* **1988**, *110*, 4356.
18. Tsuji, J. *Pure Appl. Chem.* **1989**, *61*, 1673.
19. Trost, B.M.; Kuo, GH.; Benneche, T. *J. Am. Chem. Soc.* **1988**, *110*, 621.
20. (a) Trost, B.M.; Klun, T.P. *J. Am. Chem. Soc.* **1981**, *103*, 1864. (b) Trost, B.M.; Klun, T.P. *J. Am. Chem. Soc.* **1979**, *101*, 6756. (c) Trost, B.M.; Runge, T.A. *J. Am. Chem. Soc.* **1981**, *103*, 2485, 7550, 7559.
21. Genet, J.P.; Piau, F. *J. Org. Chem.* **1981**, *46*, 2414.
22. Backvall, J.E.; Vagberg, J.O.; Granberg, K.L. *Tetrahedron Lett.* **1989**, *30*, 617.
23. van der Heide, T.A.; van der Baan, J.L.; Bickelhaupt, F.; Klumpp, G.W. *Tetrahedron Lett.* **1992**, *33*, 475.
24. Takemoto, T.; Nishikimi, Y.; Sodeoka, M.; Shibasaki, M. *Tetrahedron Lett.* **1992**, 3527; **1992**, 3531.
25. Trost, B.M.; Ohmori, M.; Boyd, S.A.; Okawara, H.; Brickner, S.J. *J. Am. Chem. Soc.* **1989**, *111*, 8281.
26. Trost, B.M.; Vos, B.A.; Brzezowski, C.M.; Martina, D.P. *Tetrahedron Lett.* **1992**, *33*, 717.
27. Trost, B.M.; Hane, J.T.; Metz, P. *Tetrahedron Lett.* **1986**, *27*, 5695.
28. Trost, B.M.; Keinan, E. *J. Org. Chem.* **1979**, *44*, 3451; *J. Am. Chem. Soc.* **1978**, *100*, 7779.
29. Bystrom, S.E.; Aslanian, R.; Backvall, J-E. *Tetrahedron Lett.* **1985**, *26*, 1749.

30. Bolitt, V.; Chaguir, B.; Sinou, D. *Tetrahedron Lett.* **1992**, *33*, 2481.
31. Gunderson, L.L.; Benneche, T.; Undheim, K. *Tetrahedron Lett.* **1992**, *33*, 1085.
32. Tenaglia, A.; Waegell, B. *Tetrahedron Lett.* **1988**, *29*, 4851.
33. Trost, B.M.; Scanlan, T.S. *J. Am. Chem. Soc.* **1989**, *111*, 4988.
34. Godleski, S.A.; Meinhart, J.D.; Miller, D.J.; Van Wallendae, S. *Tetrahedron Lett.* **1981**, *22*, 2247.
35. Trost, B.M.; Cossy, J. *J. Am. Chem. Soc.* **1982**, *104*, 6881.
36. Trost, B.M.; Tenaglia, A. *Tetrahedron Lett.* **1988**, *29*, 2927.
37. Larock, R.C.; Lee, N.H. *J. Org. Chem.* **1991**, *56*, 6253.
38. Backvall, J.E. in *Advances in Metal-Organic Chemistry*, Liebeskind, L.S.; ed. JAI Press, London, 1989, vol. 1, p. 135.
39. Backvall, J.E.; Bystrom, S.E.; Nordberg, R.E. *J. Org. Chem.* **1984**, *49*, 4619.
40. Backvall, J.E.; Nystrom, S.E.; Nordberg, R.E. *J. Am. Chem. Soc.* **1985**, *107*, 3676; 6892.
41. Backvall, J.E.; Andersson, P.G. *J. Am. Chem. Soc.* **1992**, *114*, 6374.
42. Keinan, E.; Greenspoon, N. *J. Org. Chem.* **1983**, *48*, 3545.
43. (a) Mandai, T.; Matsumoto, T.; Kawada, M.; Tsuji, J. *J. Org. Chem.* **1992**, *57*, 1326. (b) Mandai, T.; Suzuki, S.; Murakami, T.; Fujita, M.; Kawada, M.; Tsuji, J. *Tetrahedron Lett.* **1992**, *33*, 2987.
44. Hauser, F.M.; Tommasi, R.; Hewawasam, P.; Rho, Y.S. *J. Org. Chem.* **1988**, *53*, 4886.
45. Mandai, T.; Matsumoto, T.; Nakao, Y.; Teramoto, A.; Kawada, M.; Tsuji, J. *Tetrahedron Lett.* **1992**, *133*, 2549.
46. Deziel, R. *Tetrahedron Lett.* **1987**, *28*, 4371
47. Hayakawa, Y.; Kato, H.; Uchiyama, H.; Kajino, H.; Noyori, J. *Org. Chem.* **1986**, *51*, 2400.
48. Martin, S.F.; Davidsen, S.K. *J. Am. Chem. Soc.* **1984**, *106*, 6431.
49. Del Valle, L.; Stille, J.K.; Hegedus, L.S. *J. Org. Chem.* **1990**, *55*, 3019.
50. Echavarren, A.M.; Tueting, D.R.; Stille, J.K. *J. Am. Chem. Soc.* **1988**, *110*, 4039.
51. Teuting, D.R.; Echavarren, A.M.; Stille, J.K. *Tetrahedron* **1989**, *45*, 979.
52. Farina, V.; Baker, S.R.; Benigni, D.A.; Sapino, C., Jr. *Tetrahedron Lett.* **1988**, *29*, 5739.
53. Katsamura, S.; Fujiwara, S.; Isoe, S. *Tetrahedron Lett.* **1987**, *28*, 1191.
54. vander Louw, J.; vander Baan, J.L.; deKanter, F.J.J.; Bickelhaupt, F.; Klumpp, G.W. *Tetrahedron* **1992**, *48*, 6087.
55. For reviews see: (a) Oppolzer, W. *Angew. Chem. Int. Ed. Engl.* **1989**, *28*, 38. (b) Oppolzer, W. *Pure Appl. Chem.* **1990**, *62*, 1941.
56. Oppolzer, W.; DeVita, R.J. *J. Org. Chem.* **1991**, *56*, 6256.
57. Grigg, R.; Sridharan, V.; Surkirthalingam, S. *Tetrahedron Lett.* **1991**, *32*, 3855.
58. Trost, B.M.; Luengo, J.I. *J. Am. Chem. Soc.* **1988**, *110*, 8239.
59. Harris, G.D., Jr.; Herr, R.J.; Weinreb, S.M. *J. Org. Chem.* **1992**, *57*, 2528.
60. Oppolzer, W.; Keller, T.H.; Bedoya-Zurita, M.; Stone, C. *Tetrahedron Lett.* **1989**, *30*, 5883. For related processes see: (a) Oppolzer, W.; Gaudin, J.M.; Birkinshaw, T.N. *Tetrahedron Lett.* **1988**, *29*, 4705. (b) Oppolzer, W.; Gaudin, J.M. *Helv. Chim. Acta* **1987**, *70*, 1477.
61. Oppolzer, W.; Bienayme, H.; Genevois-Borella, A. *J. Am. Chem. Soc.* **1991**, *113*, 9660.
62. Oppolzer, W.; Xu, J-Z.; Stone, C. *Helv. Chim. Acta* **1991**, *74*, 465.
63. Keese, R.; Guidetti-Grept, R.; Herzog, B. *Tetrahedron Lett.* **1992**, *33*, 1207.
64. For reviews see: (a) Trost, B.M. *Angew. Chem. Int. Ed. Engl.* **1986**, *25*, 1.(b) Trost, B.M. *Pure Appl. Chem.* **1988**, *60*, 1615.

65. (a) Trost, B.M.; Lynch, J.; Renaut, P.; Steinman, D.H. *J. Am. Chem. Soc.* **1986**, *108*, 284. (b) Trost, B.M.; Mignani, S.M. *Tetrahedron Lett.* **1986**, *27*, 4137.
66. Trost, B.M.; Nanninga, T.N.; Satoh, T. *J. Am. Chem. Soc.* **1985**, *107*, 721.
67. Trost, B.M.; Seoane, P.R. *J. Am. Chem. Soc.* **1987**, *109*, 615.
68. Trost, B.M.; Grese, T.A.; Chan, D.M.T. *J. Am. Chem. Soc.* **1991**, *113*, 7350.
69. Trost, B.M.; Grese, T.A. *J. Am. Chem. Soc.* **1991**, *113*, 7363.
70. Trost, B.M.; Grese, T.A. *J. Org. Chem.* **1992**, *57*, 686.
71. Trost, B.M.; King, S.A. *J. Am. Chem. Soc.* **1990**, *112*, 408.
72. Trost, B.M.; Sharma, S.; Schmidt, T. *J. Am. Chem. Soc.* **1992**, *114*, 7903.
73. Binger, P.; Büch, H.M. *Top. Curr. Chem.* **1987**, *135*, 77.
74. Lewis, R.T.; Motherwell, W.B.; Shipman, M. *J. Chem. Soc., Chem. Comm.* **1988**, 948.
75. Trost, B.M.; Hung, M-H. *J. Am. Chem. Soc.* **1984**, *106*, 6837.
76. Trost, B.M.; Merlic, C.A. *J. Am. Chem. Soc.* **1990**, *112*, 9590.
77. Vong, W-J.; Peng, S-M.; Lin, S-H.; Lin, W-J.; Liu, R-S. *J. Am. Chem. Soc.* **1991**, *113*, 573.
78. Roustan, J.L.; Merour, J.Y.; Houlihan, F. *Tetrahedron Lett.* **1979**, 3721.
79. Li, Z.; Nicholas, K.M. *J. Organomet. Chem.* **1991**, *402*, 105.
80. Green, J.; Carroll, M.K. *Tetrahedron Lett.* **1991**, *32*, 1141.
81. Heck, R.F.; Breslow, D.S. *J. Am. Chem. Soc.* **1963**, *85*, 2779.
82. Heck, R.F. in *Organic Synthesis via Metal Carbonyls*, Wender, P. and Pino, P., eds. Wiley, New York, 1968, vol. 1, pp. 379-384.
83. (a) Hegedus, L.S.; Inoue, Y. *J. Am. Chem. Soc.* **1982**, *104*, 4917. (b) Hegedus, L.S.; Perry, R.J. *J. Org. Chem.* **1984**, *49*, 2570.
84. For reviews, see: (a) Semmelhack, M.F. *Org. React.* **1972**, *19*, 115. (b) Billington, D.C. *Chem. Soc. Rev.* **1985**, *14*, 93.
85. Corey, E.J.; Semmelhack, M.F. *J. Am. Chem. Soc.* **1967**, *89*, 2755.
86. (a) Hegedus, L.S.; Stiverson, R. *J. Am. Chem. Soc.* **1974**, *96*, 3250. (b) Guerrieri, F.; Chinsoli, G.P.; Merzoni, S. *Gazz. Chim. Ital.* **1974**, *104*, 557.
87. Saito, K.; Inoue, S.; Sato, K. *J. Chem. Soc., Perkin I* **1973**, 2289.
88. Korte, D.E.; Hegedus, L.S.; Wirth, R.K. *J. Org. Chem.* **1977**, *42*, 1329.
89. Hegedus, L.S.; Sestrick, M.R.; Michaelson, E.T.; Harrington, P.J. *J. Org. Chem.* **1989**, *54*, 4141.
90. Johnson, J.R.; Tully, P.S.; Mackenzie, P.B.; Sabat, M. *J. Am. Chem. Soc.* **1991**, *113*, 6172.
91. Grisso, B.A.; Johnson, J.R.; Mackenzie, P.B. *J. Am. Chem. Soc.* **1992**, *114*, 5160.
92. Hegedus, L.S.; Thompson, D.H.P. *J. Am. Chem. Soc.* **1985**, *107*, 5663.
93. Hegedus, L.S.; Wagner, S.D.; Waterman, E.L.; Siirala-Hansen, K. *J. Org. Chem.* **1975**, *40*, 593.

CHAPTER 10

Synthetic Applications of Transition Metal Arene Complexes

10.1 Introduction

Several transition metals, including Cr, Mo, W, Fe, Ru, V, Mn, and Rh, form stable, isolable arene complexes, and a few of these have found use in organic synthesis. Arene complexes of chromium have been most extensively developed, because they are easy-to-prepare, air-stable yellow-to-red solids. They undergo a number of useful organic reactions, and the arene group is readily released from complexation by mild oxidation (air/light, I_2, $FeCl_3$ Ce(IV)), making recovery of the metal-free organic compound simple and convenient.

Arenechromium tricarbonyl complexes are normally prepared by heating $Cr(CO)_6$ in the arene as solvent, although the procedure is complicated by the sublimation of $Cr(CO)_6$.[1] Alternatively, one can start with preformed amine complexes such as $Cr(CO)_3(NH_3)_3$ or $Cr(CO)_5(2-picoline)$ to avoid this problem (Eq. 10.1).[2] Perhaps the most convenient method is to carry out an arene exchange with the (naphthalene) $Cr(CO)_3$ complex (Eq. 10.2),[3] a procedure which goes under mild conditions and does not require large excesses of arene.

Eq. 10.1

$$M(0),\, d^6,\, 18e^-,\, \text{sat.}$$

Eq. 10.2

$$(\text{Naphthalene})\text{Cr(CO)}_3 \;+\; \text{Ar}'\text{H} \;\rightleftharpoons\; \text{Ar}'\text{H Cr(CO)}_3 \;+\; \text{Naphthalene}$$

The complexation of arenes to the chromium tricarbonyl fragment is facilitated by electron donating groups on the arene ring, and, in general, electron rich arenes readily form arenechromium tricarbonyl complexes, while electron poor arenes react much more slowly or not at all. (The chromium tricarbonyl complex of nitrobenzene is unknown.) In some cases, discrimination between two similar arene rings in the same compound is possible, provided there is sufficient electronic difference between them (see Eqs. 10.17 and 10.30 for examples).

Complexation of arenes has a profound effect on their reaction chemistry (Figure 10.1).[4] Relative to the arene ring, the Cr(CO)₃ fragment is net electron withdrawing, as evidenced by the high dipole moment (5.08D for benzene chromium tricarbonyl), the increase in acidity of benzoic acid upon complexation ($pK_a = 4.77$ for the benzoic acid complex vs. 5.75 for free benzoic acid) and the decrease in basicity for complexed aniline ($pK_b = 13.31$ vs 11.70 for aniline itself). As a consequence, the arene ring becomes activated towards nucleophilic attack, rather than the normal electrophilic attack. In addition, the electron deficient arene ring is better able to stabilize negative charge, thus both ring and benzylic deprotonation become more favorable. Finally, the chromium tricarbonyl fragment completely blocks one face of the arene, and directs incoming reagents to the face opposite the metal. All of these effects have been used to an advantage in organic synthesis.

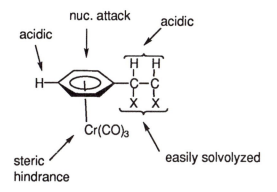

Figure 10.1   Effects of Complexation of Arenes to Chromium

10.2 Reactions of Arenechromium Tricarbonyl Complexes[5]

a. Nucleophilic Aromatic Substitution of Aryl Halides

Although simple aryl halides are relatively inert to nucleophilic aromatic substitution, when complexed to chromium, this reaction is greatly facilitated. For example, the rate of substitution of chloride by methoxide in η^6-chlorobenzenechromium tricarbonyl approximates that of uncomplexed nitrobenzene,[6] an indication of the electron withdrawing power of the chromium tricarbonyl fragment. Sodium phenoxide and aniline also readily effect this substitution.[7] η^6-Fluorobenzenechromium tricarbonyl is even more reactive, undergoing substitution by a range of nucleophiles including alkoxides, amines, cyanide[8] and stabilized carbanions (Eq. 10.3).[9] This chemistry was used to synthesize α-methyl phenyl glycine (Eq. 10.4).[10]

Eq. 10.3

X = Ph, CO$_2$Et
Y = CN, CO$_2$Et

70-80%

Eq. 10.4

These substitution reactions all proceed by a two-step process, involving nucleophilic attack of the arene ring from the face opposite the metal, forming an anionic η^5-cyclohexadienyl complex, followed by rate-limiting loss of the halide from the endo side of the ring (Eq. 10.5). The displacement of halide by carbanions is limited to stabilized carbanions capable of adding reversibly to the complexed arene. More reactive carbanions, such as 2-lithio-1,3-dithiane, attack the complexed arene *ortho*, and *meta* to the halide, producing η^5-cyclohexadienyl complexes which cannot directly lose chloride, and furthermore, cannot rearrange (equilibrate) to the η^5-cyclohexadienyl complex, which can (Eq. 10.5).[11] Stabilized carbanions also initially attack complexed chlorobenzene *ortho* and *meta* to the chloride, but the addition is reversible, and, upon equilibration, eventually attack at the halide-bearing position occurs, followed by loss of halide, resulting in substitution.

Eq. 10.5

b. Addition of Carbon Nucleophiles to Arenechromium Complexes[12]

A range of stabilized carbanions attack chromium-complexed arenes from the face opposite the metal, producing anionic η^5-cyclohexadienyl complexes (Eq. 10.6). With substituted aromatics, under conditions of kinetic control, the regioselectivity is that expected for nucleophilic aromatic substitution (*meta* to electron-donating groups, *para* to acceptor groups) and predicted by the assumption that the frontier molecular orbitals are arene-centered.[13a,b] Ortho addition occurs with groups capable of coordinating the incoming nucleophile (oxazolines, imines). In contrast to benzeneCr(CO)3 complexes, for which MeLi and n-BuLi react by deprotonation, these reagents *add* to phenyloxazoline and phenylmethaneimine complexes.[13c] As might be expected for *nucleophilic* aromatic substitution, replacement of an electron withdrawing CO ligand with a ligand such as phosphine, which increases the electron density on the metal, decreases the overall reactivity of arene complexes toward nucleophilic attack.

Eq. 10.6

Very reactive carbanions, such as dithiane lithium, phenyllithium or anisole lithium, add irreversibly to the complexed arene. The resulting anionic η^5-cyclohexadienyl complexes are relatively stable and can sometimes be isolated. Oxidation of these complexes generates the alkylated arene, protonolysis generates mixtures of isomeric cyclohexadienes, and reaction with carbon electrophiles produces trans disubstituted species (see below). In contrast, stabilized carbanions add to complexed arenes reversibly, and are always in equilibrium with the starting compounds. (This dissociation of the carbanion can be largely suppressed by the use of a polar cosolvent (HMPA, DMPU) or by using the K rather than the Li C-nucleophile).[13d] In THF, oxidative cleavage again produces the alkylated arene, and protonolysis the cyclohexadienes, but reaction with *carbon* electrophiles results in regeneration of the starting arene complex and *alkylation* of the electrophile by the stabilized carbanion (Eq. 10.6).

Both the activating and the directing effects of complexation to chromium are synthetically useful. For example, both indoles and benzofurans are electron rich aromatic compounds normally inert to nucleophilic substitutions, but are reactive to electrophilic substitution at the 2 and 3 positions. However, complexation to chromium permits nucleophilic alkylation to take place, and predominantly at the normally unaccessible C-4 position or, if the C-3 substituent is large, at the C-7 position (Eq. 10.7).[14]

Eq. 10.7

major   minor

X = NH, NMe, NBz, O

R⁻ = ⁻CH₂CN, ...

Protolytic cleavage of the η^5-cyclohexadienyl complexes resulting from alkylation of η^6-arenechromium tricarbonyl complexes produces cyclohexadienes rather than arenes (Eq. 10.8).[15] With anisole as substrate this is a useful procedure for the synthesis of substituted cyclohexenones. This ability of chromium to activate arenes towards nucleophilic attack, to direct the attack to specific positions on the arene ring, and to permit protolytic cleavage of the η^5-cyclohexadienyl complex to produce cyclo-hexenones has been used in a notewworthy total synthesis of acorenone B (Eq. 10.9).[16] Again, protons are the only electrophile that cleaves the η^5-cyclohexadienyl complex. Since most alkylations (see exceptions below) are reversible, other electrophiles react preferentially with the free carbanion, regenerating the η^6-arene chromium tricarbonyl complex, and alkylating the electrophile.

Eq. 10.8

Eq. 10.9

(±)-Acorenone B

The reversibility of alkylation by most carbanions is clearly demonstrated by the chemistry of the η^6-(1,4-dimethoxynaphthalene) chromium tricarbonyl complex, which undergoes kinetic alkylation at β-

position, but thermodynamic alkylation at the α-position (Eq. 10.10).[17] In contrast to stabilized carbanions, the dithiane anion adds irreversibly to η⁶-arenechromium complexes, resulting in exclusive production of the kinetic product. The differences among various carbanions is clearly seen in Eq. 10.11.[18] The kinetic alkylation of complexed dimethoxynaphthalene was used in a synthesis of a daunomycin intermediate (Eq. 10.12).[17]

Eq. 10.10

Eq. 10.11

R = CH₂CO₂tBu	1:4	66%
CH₂CN	1:3	86%
$\rangle$—CN	1:2.6	68%
dithiane	100:0	86%

Eq. 10.12

Since dithiane anion adds irreversibly to the η^6-arenechromium tri-carbonyl complex, it should, in principle, be possible to trap the resulting anionic η^5-cyclohexadienyl complex with electrophiles other than a pro-ton. Indeed, treatment of this complex with reactive alkyl halides does result in a clean reaction but, surprisingly, an *acyl* group rather than the corresponding alkyl group is introduced, and from the same face as the metal, resulting in clean *trans* difunctionalization of the initial arene (Eq. 10.13).[19] To account for this, initial nucleophilic attack by the dithiane had to occur from the face opposite the metal, as expected.

Eq. 10.13

This is followed by alkylation at the *metal*, then migration to an adjacent carbonyl group to produce an acyl intermediate. Reductive elimination from this complex would deliver the acyl group from the *same* face as the metal, resulting in the clean *trans* disubstitution observed. This chemistry works equally well with substituted naphthalenes (Eq. 10.14).[20a] This one-pot addition of two carbon substituents across an arene double bond serves as the key step in a synthesis of the aklavinone AB ring (Eq. 10.15).[20a] Combining this methodology with a chiral oxazoline auxiliary results in highly diastereoselective syntheses of interesting products (Eq. 10.16). Note that benzyl and allyl groups are transferred from the metal without CO insertion, a consequence of their lower aptitude towards migratory CO insertion.[20b,c]

Eq. 10.14

Eq. 10.15

Eq. 10.16

c. Lithiation of Arenechromium Tricarbonyl Complexes

Complexation of an arene to chromium activates the ring towards lithiation, and ring-lithiated arenechromium tricarbonyl complexes are readily prepared, and generally reactive towards electrophiles (Eq. 10.17).[21] With arenes having substituents with lone pairs of electrons, such as MeO, F, or Cl, lithiation always occurs *ortho* to the substituted position. This process is efficient even with quite complex systems, and high selectivity can be achieved. For example, treatment of dihydrocryptopine with chromium hexacarbonyl resulted in exclusive complexation of the slightly more electron rich dimethoxy benzene ring, perhaps with some assistance from the benzylic hydroxy group (see Eq. 10.32). Lithiation occurred exclusively *ortho* to the methoxy group and peri to the benzylic hydroxy group, resulting in regiospecific alkylation (Eq. 10.18).[22]

Eq. 10.17

Y = H, OMe, F, Cl
E$^+$ = CO$_2$, MeI, PhCHO, TMSCl,

Eq. 10.18

82%

(1) BuLi
(2) MeI

98%

Complexed indoles are lithiated in the normally unreactive 4-position provided a large protecting group is present on nitrogen to suppress lithiation at the 2-position (Eq. 10.19).[23]

Eq. 10.19

E = TMS, CO$_2$Et, CO$_2$Me,

Because η^6-arenechromium tricarbonyl complexes are the products of ring lithiation/electrophilic attack, nucleophilic attack on the complexed ring is still possible (Eq. 10.20),[24] as in the synthesis of frenolicin. Note the palladium-catalyzed alkoxycarbonylation (Chapter 7) in the later stages of this synthesis. By coupling o-lithiation with the facile displacement of fluoride from complexed fluorobenzene, a variety of heterocycles are readily accessible (Eq. 10.21[25] and Eq. 10.22[26]).

Eq. 10.20

Frenolicin

Eq. 10.21

Eq. 10.22

d. Activation of the Benzylic Position of Arenechromium Tricarbonyl Complexes[27]

Another manifestation of the "electron-withdrawing" properties of the $Cr(CO)_3$ group when complexed to aromatic compounds is its ability to stabilize negative charge at the benzylic position of alkyl side chains. Thus, ethyl benzene underwent facile α-alkylation to produce i-propylbenzene when complexed to chromium (Eq. 10.23).[28] Complexed allylbenzene underwent α-deprotonation followed by γ-alkylation (Eq. 10.24).[29] Complexed styrenes undergo alkylation at the β-position, generating the stabilized benzyl anion, followed by α-alkylation by electrophiles (Eq. 10.25).[30] In combination with other chromium-based arene chemistry, quite complex molecules can be assembled quickly (Eq. 10.26).[31]

Eq. 10.23

Eq. 10.24

Eq. 10.25

R^1 = H, Me, SEt

EX = HX, MeI, PhCOCl

R^2 = CMe_2CN, CMe_2CO_2tBu,
    C(CN)(Me)(OR), nBu, Ph,

Most remarkably, chromium also stabilizes *positive* charge in the benzylic position of complexed arenes, clearly not a function of the electron-withdrawing ability of the Cr(CO)$_3$ fragment. Instead, it is thought that complexed benzylic cations are stabilized by a "through space" interaction with the metal-bound carbonyl groups and/or filled metal *d*-orbitals. Whatever the mechanism of stabilization, the phenomenon permits useful synthetic transformations via the stabilized benzyl cation (Eq. 10.27[32] and Eq. 10.28[33]).

Eq. 10.27

Eq. 10.28

e. Steric Effects with Arenechromium Tricarbonyl Complexes

In arenechromium tricarbonyl complexes, the chromium tricarbonyl fragment completely blocks one face of the arene, and all of the reactions discussed above (except the acylation reaction in Eqs. 10.13 and 10.14) occur from the face opposite the metal. This can result in a very high degree of stereocontrol, and has found many uses in organic synthesis. A very early example of this is seen in Eq. 10.29,[34] in which α-alkylation of an enone by complexed α-methyl indanone occurred exclusively from the face opposite the metal. Removal of an activated benzylic proton permitted intramolecular alkylation of the keto group, again from the face opposite the metal, resulting in clean *cis* stereochemistry. Similarly, complexed steroids undergo benzylic reactions from the face opposite the metal (Eq. 10.30).[35] Complexed tetrahydroprotoberberins could also be stereospecifically methylated from the face opposite the metal, as well as ring alkylated (Eq. 10.31).[36] The initial complex was formed in the usual manner, by heating the alkaloid and $Cr(CO)_6$ together. Complexation occurred exclusively at the more electron rich dimethoxy (rather than the dioxomethylene) ring, but with no facial selectivity. (The same lack of facial selectivity was observed in Eq. 10.30.) However, the diastereoisomers were easily separated. The one shown in Eq. 10.31 underwent clean ring lithiation/methylation α to the methoxy group, then regioselective benzylic methylation α- to nitrogen from the face opposite the metal.

Eq. 10.29

Eq. 10.30

Eq. 10.31

In contrast to the two cases above, racemic 2-indanol undergoes facially specific complexation, from the same face as the OH group, presumably directed there by complexation to the oxygen. These enantiomeric complexes can be resolved, and oxidation of one of the enantiomers produces an indanone complex which is optically active solely by virtue of the face of the arene occupied by the Cr(CO)$_3$ fragment (see below). Since reactions of this optically active complex occur exclusively from the face opposite the metal, alkylation by Grignard reagents produces a single enantiomer of the indanol, α-alkylation a single enantiomer of the indanone, and α-alkylation followed by reduction a single enantiomer of the α-alkyl indanol (Eq. 10.32).[37] Similar transformations are feasible with tetralone.[38]

Eq. 10.32

As implied above, unsymmetrically 1,2- and 1,3-disubstituted arenechromium tricarbonyl complexes are chiral and are enantiomeric on the basis of which face of the arene the chromium tricarbonyl fragment occupies (Figure 10.3). These can often be resolved,[39] or in some cases prepared by asymmetric synthesis.[40] Since reactions occur from the face opposite the metal, highly enantioselective reactions can be achieved (Eq. 10.33[41] and Eq. 10.34[42]).

Eq. 10.33

Eq. 10.34

Planar chiral areneCr(CO)$_3$ complexes can also be generated by diastereoselective reactions. For example, α-lithiation of the Cr(CO)$_3$ complex of (R)-N,N-dimethyl-α-phenethyl amine occurs almost exclusively at one of the two diastereotopic ortho positions, and the resulting carbanion alkylates aldehydes almost exclusively from a single face (Eq. 10.35).[43] This selectivity is a result of amine directed lithiation from the less hindered conformer (methyl *anti* to the large Cr(CO)$_3$ fragment) and subsequent alkylation of the aldehyde from the face having the aldehyde R group *anti* to the bulky Cr(CO)$_3$ fragment. Alternatively, diastereoselective nucleophilic addition ortho to a chiral hydrazone group followed by hydride abstraction and removal of the chiral auxiliary gives the planar chiral benzaldehyde complexes with high enantiomeric excess (Eq. 10.36).[44] Since all classes of reactions of arenechromium tricarbonyl complexes are subject to steric control by the metal, many useful synthetic transformations are available. The steric effect of the Cr(CO)$_3$ group is

also demonstrated in the cycloaddition reaction shown in Eq. 10.37.[45] Anion accelerated ring opening of the benzocyclobutene generates an o-quin-
odimethane complex. As the Cr(CO)$_3$ group blocks one enantioface, the
cycloaddition reaction takes place exclusively from the side opposite the
organometallic group. A fine example of the multiple uses chromium carbonyl can have in aromatic chemistry is seen in Eq. 10.38,[46] in which complexation to chromium is used first to direct a reduction, then to stabilize
a benzyl cation and direct a benzylic alkylation, then to direct a methyllithium reaction with a ketone, again to direct a reduction of an alcohol,
and finally to activate the arene ring toward nucleophilic attack!

Eq. 10.35

vs.

>97% de

Eq. 10.36

>93% ee

Eq. 10.37

Eq. 10.38

10.3 Reactions of Other Arenemetal Complexes

By far the most research involving metal arene complexes in synthesis has focused on chromium complexes. However, a number of other metal arene complexes are potentially useful and are attracting more attention. Cationic cyclopentadienyliron complexes of substituted arenes undergo ortho alkylation by carbanions to the substituent (Eq. 10.39).[47] Oxidation by DDQ removes the metal to give the alkylated arene in good yield. Iron complexed tetralin undergoes both ring alkylation and benzylic alkylation with a high degree of stereocontrol (Eq. 10.40).[48] When the nucleophile is phenoxide, iron complexed chlorobenzenes undergo substitution reactions to form aryl ethers (Eq. 10.41).[49] The corresponding ruthenium complexes behave similarly (Eq. 10.42[50], Eq. 10.43[51], and Eq. 10.44[52]).

Eq. 10.39

Nuc. = CN^-, $\bar{C}H_2CO_2Me$

X = Cl, CN, NO_2, COPh

Eq. 10.40

Eq. 10.41

40-90%

Eq. 10.42

Nuc = MeO⁻, PhCH₂O⁻, Me₂NH,

Eq. 10.43

60-90%

M = Fe, Ru

Nuc = pMeOPhO⁻, mNO₂PhO⁻, O⟩NH ,

Nuc′ = Nuc as well as ⁻CH(CO₂R)

Eq. 10.44

60-80%

Cationic arenemanganese tricarbonyl complexes are similarly reactive toward nucleophiles and can be alkylated (Eq. 10.45[53] and Eq. 10.46[54]) and alkoxylated (Eq. 10.47).[55] Again, the metal must be removed by oxidation to free the organic ligand.

Eq. 10.45

R = H, OMe, OPh
R¹ = H, OMe, OPh

75%   9:1

Eq. 10.46

85%

$$\xrightarrow[\text{(ox)}]{\text{NBS}} \quad \xrightarrow[\text{HCl}]{0.25\ \text{M}}$$

70% ee

Eq. 10.47

60%

References

1. (a) Nicholls, B.; Whiting, M.C. *J. Chem. Soc.* **1959**, 551. (b) Mahaffy, C.A.L.; Pauson, P. *Inorg. Synth.* **1979**, *XIX*, 154.
2. (a) Mosher, G.A.; Rausch, M.D. *Synth. React. Inorg. Metal. Org. Chem.* **1979**, *4*, 38. (b) Rausch, M.D. *J. Org. Chem.* **1974**, *39*, 1787.
3. Kundig, E.P.; Perret, C.; Spichiger, S.; Bernardinelli, G. *J. Organomet. Chem.* **1985**, *286*, 183.
4. This figure was adopted from an excellent (dated) review: Semmelhack, M.F. *J. Organomet. Chem. Libr.* **1976**, *1*, 361.
5. For a review see: (a) Kalinin, V.N. *Russ. Chem. Rev.* **1987**, *56*, 682. (b) Uemura, M. *Adv. Met. Org. Chem.* **1991**, *2*, 195.
6. Brown, D.A.; Raju, J.R. *J. Chem. Soc. A* **1966**, *40*, 1617.
7. Rosca, S.J.; Rosca, S. *Rev. Chim.* **1974**, *25*, 461.
8. Mahaffy, C.A.L.; Pauson, P.L. *J. Chem. Res.* **1979**, 128.
9. Baldoli, C.; DelButtero, P.; Licandro, E.; Maiorana, S. *Gazz. Chim. Ital.* **1988**, *118*, 409.
10. Chaari, M.; Jenhi, A.; Lavergne, J.P.; Viallefont, P. *J. Organomet. Chem.* **1991**, *401*, C10.
11. Semmelhack, M.F.; Hall, H.T., Jr. *J. Am. Chem. Soc.* **1974**, *96*, 7091, 7092.
12. For a review see: Semmelhack, M.F. in "Comprehensive Organic Synthesis", Trost, B.M., Ed., Pergamon Press, 1991, Vol. 4, pp. 517-29.
13. (a) Semmelhack, M.F.; Garcia, J.L.; Cortes, D.; Farina, R.; Hong, R.; Carpenter, B.K.; *Organometallics* **1983**, *2*, 467. (b) Semmelhack, M.F.; Seufert, W.; Keller, L. *J. Organomet. Chem.* **1982**, *226*, 183. (c) Kündig, E.P.; Amurrio, D.; Liu, R.; Ripa, A. *Synlett* **1991**, 657. (d) Kündig, E.P.; Desobry, V.; Simmons, D.P.; Wenger, E. *J. Am. Chem. Soc.* **1989**, *111*, 1804.
14. Semmelhack, M.F.; Wulff, W.; Garcia, J.L. *J. Organomet. Chem.* **1982**, *240*, C5.
15. (a) Semmelhack, M.F.; Harrison, J.J.; Thebtaranonth, Y. *J. Org. Chem.* **1979**, *44*, 3275. (b) Boutonnet, J.C.; Levisalles, J.; Normant, J.M.; Rose, E. *J. Organomet. Chem.* **1983**, *255*, C21.
16. Semmelhack, M.F.; Yamashita, A. *J. Am. Chem. Soc.* **1980**, *102*, 5924.
17. Kündig, E.P.; Desobry, V.; Simmons, D.P. *J. Am. Chem. Soc.* **1983**, *105*, 6962.
18. Cambie, R.C.; Clark, G.R.; Gallagher, S.R.; Rutledge, R.S.; Stone, M.J.; Woodgate, P.D. *J. Organomet. Chem.* **1988**, *342*, 315, 317.
19. Kündig, E.P.; Cunningham, A.F., Jr.; Paglia, P.; Simmons, D.P.; Bernardinelli, G. *Helv. Chim. Acta* **1990**, *73*, 386.
20. (a) Kündig, E.P.; Inage, M.; Bernardinelli, G. *Organometallics* **1991**, *10*, 2921. (b) Kündig, E.P.; Bernardinelli, G.; Liu, R.; Ripa, A. *J. Am. Chem. Soc.* **1991**, *113*, 9676. (c) Kündig, E.P.; Ripa, A.; Bernardinelli, G. *Angew. Chem. Int. Ed.*, **1992**, *31*, 1071.
21. (a) Semmelhack, M.F.; Bisaha, J.; Czarny, M. *J. Am. Chem. Soc.* **1979**, *101*, 768. (b) Card, R.J.; Trahanovsky *J. Org. Chem.* **1980**, *45*, 2560. (c) Gilday, J.P.; Negri, J.T.; Widdowson, D.A. *Tetrahedron* **1989**, *45*, 4605. (d) Dickens, P.J.; Gilday, J.P.; Negri, J.T.; Widdowson, D.A. *Pure Appl. Chem.* **1990**, *62*, 575. (e) Kündig, E.P.; Perret, C.; Rudolph, B. *Helv. Chim. Acta* **1990**, *73*, 1970.
22. Davies, S.G.; Goodfellow, C.L.; Peach, J.M.; Waller, A. *J. Chem. Soc., Perkin I* **1991**, 1019.
23. (a) Nechvatal, G.; Widdowson, D.A. *J. Chem. Soc., Chem. Comm.* **1982**, 467. (b) Beswick, P.J.; Greenwood, C.S.; Mowlem, T.J.; Nechvatal, G.; Widdowson, D.A. *Tetrahedron* **1988**, *44*, 7325. (c) Masters, N.F.; Mathews, N.; Nechvatal, G.; Widdowson, D.A. *Tetrahedron* **1989**, *45*, 5955. (d) Dickens, M.J.; Mowlen, T.J.;

Widdowson, D.A.; Slawin, A.M.Z.; Williams, D.J. *J. Chem. Soc., Perkin Trans I* **1992**, 323.

24. Semmelhack, M.F.; Zask, A. *J. Am. Chem. Soc.* **1983**, *105*, 2034.

25. Ghavshou, M.; Widdowson, D.A. *J. Chem. Soc., Perkin I* **1983**, 3065.

26. Dickens, P.J.; Slawin, A.M.Z.; Widdowson, D.A.; Williams, D.J. *Tetrahedron Lett.* **1988**, *29*, 103.

27. For reviews see: (a) Senechal-Torquer, M.C.; Senechal, D.; LeBihan, J.Y.; Gentric, D.; Caro, B. *Bull. Soc. Chim. Fr.* **1992**, *129*, 1121. (b) Davies, S.G.; Coote, S.J.; Goodfellow, C.L. *Adv. Met.-Org. Chem.* **1991**, *2*, 1.

28. Jaouen, G.; Meyer, A.; Simmoneaux, G. *J. Chem. Soc., Chem. Comm.* **1975**, 813.

29. (a) Gentric, D.; LeBihan, J-Y.; Senechal-Toquer, M.C.; Senchal, D.; Caro, B. *Tetrahedron Lett.* **1986**, *27*, 3849. (b) Senechal-Tocquer, M.C.; Senechal, D.; LeBihan, J-Y.; Gentric, D.; Caro, B. *J. Organomet. Chem.* **1991**, *420*, 185.

30. Semmelhack, M.F.; Suefert, W.; Keller, L. *J. Am. Chem. Soc.* **1980**, *102*, 6584.

31. Uemura, M.; Minami, T.; Hayashi, Y. *J. Chem. Soc., Chem. Comm.* **1984**, 1193.

32. Uemura, M.; Minami, T.; Hayashi, Y. *J. Organomet. Chem.* **1986**, *299*, 119.

33. (a) Coote, S.J.; Davies, S.G.; Middlemiss, D.; Naylor, A. *Tet. Asymm.* **1990**, *1*, 33. (b) Davies, S.G.; Donohoe, T.J.; Williams, J.M.J. *Pure Appl. Chem.* **1992**, *64*, 379.

34. Jaouen, G.; Meyer, A. *Tetrahedron Lett.* **1976**, 3547.

35. Jaouen, G.; Top, S.; Laconi, A.; Conturier, D.; Brocard, J. *J. Am. Chem. Soc.* **1984**, *106*, 2207.

36. (a) Blagg, J.; Davies, S.G. *J. Chem. Soc., Chem. Comm.* **1986**, 492. (b) Davies, S.G.; Goodfellow, C.L.; Peach, J.M.; Waller, A. *J. Chem. Soc., Perkin I* **1991**, 1019.

37. Meyer, A.; Jaouen, G. *J. Chem. Soc., Chem. Comm.* **1974**, 787.

38. Schmalz, H.G.; Millies, B.; Bats, J.W.; Dürner, G. *Angew. Chem. Int. Ed. Engl.* **1992**, *31*, 631.

39. (a) Solladie-Cavallo, A.; Solladie, G.; Tsamo, E. *J. Org. Chem.* **1979**, *44*, 4189. (b) Davies, S.G.; Goodfellow, C.L. *J. Chem. Soc. Perkin Trans I* **1989**, 192.

40. (a) Kondo, Y.; Green, J.R.; Ho, J. *J. Org. Chem.* **1991**, *56*, 7199. (b) Alexakis, A.; Mangeney, P.; Marck, I.; Rose-Munch, F.; Rose, E.; Semra, A.; Robert, F. *J. Am. Chem. Soc.* **1992**, *114*, 8288.

41. Davies, S.G.; Goodfellow, C.L. *J. Chem. Soc., Perkin I* **1990**, 393.

42. Colonna, S.; Manfredi, A.; Solladie-Cavallo, A.; Quazzotti, S. *Tetrahedron Lett.* **1990**, *31*, 6185.

43. (a) Uemura, M.; Miyake, R.; Shiro, M.; Hayashi, Y. *Tetrahedron Lett.* **1991**, *32*, 4569. (b) Blagg, J.; Davies, S.G.; Goodfellow, C.L.; Sutton, K.H. *J. Chem. Soc. Perkin Trans I* **1987**, 1805. (c) Heppert, J.A. Aube, J.; Thomas-Miller, M.E.; Milligen, M.L.; Takusagawa, F. *Organometallics* **1990**, *9*, 727.

44. Kündig, E.P.; Liu, R.; Ripa, A. *Helv. Chim. Acta* **1992**, *79*, 2657.

45. (a) Kündig, E.P.; Bernardinelli, G.; Leresche, J. *J. Chem. Soc., Chem. Commun.* **1991**, 1713. (b) Wey, H.G.; Butenschön, H. *Angew. Chem. Int. Ed.* **1991**, *30*, 880.

46. Uemura, M.; Nishmura, H.; Minami, T.; Hayashi, Y. *J. Am. Chem. Soc.* **1991**, *113*, 5402.

47. (a) For a review see: Astruc, D. *Tetrahedron* **1983**, *39*, 4027. (b) Sutherland, R.G.; Chowdhury, R.L.; Piorko, A.; Lee, C.C. *J. Org. Chem.* **1987**, *52*, 4618. (c) Sutherland, R.G.; Zhang, C.; Piorko, A. *J. Organomet. Chem.* **1991**, *419*, 357.

48. (a) Cambie, R.C.; Jannssen, S.J.; Rutledge, P.S.; Woodgate, P.D. *J. Organomet. Chem.* **1989**, *359*, C14. (b) Grundy, S.L.; Sam, A.R.H.; Stobart, S.R. *J. Chem. Soc. Perkin I* **1989**, 1663.

49. Cambie, R.C.; Jannssen, R.J.; Rutledge, R.S.; Woodgate, P.D. *J. Organomet. Chem.* **1991**, *420*, 387.

50. (a) Moriarty, R.M.; Ku, Y-Y.; Gill, U.S. *J. Chem. Soc., Chem. Comm.* **1987**, 1493. (b) Moriarty, R.M.; Ku, Y-Y.; Guo, L. *J. Chem. Soc., Chem. Comm.* **1988**, 1621. (c) Moriarty, R.M.; Ku, Y-Y.; Gill, U.S. *Organometallics* **1988**, *7*, 660.

51. Pearson, A.J.; Park, J.G.; Zhu, P.Y. *J. Org. Chem.* **1992**, *57*, 3583.

52. Pearson, A.J.; Park, J.G. *J. Org. Chem.* **1992**, *57*, 1744.

53. (a) Miles, W.H.; Smiley, P.M.; Brinkman, H.R. *J. Chem. Soc., Chem. Comm.* **1989**, 1897. (b) Miles, W.H.; Brinkman, H.R. *Tetrahedron Lett.* **1992**, *33*, 589.

54. Pearson, A.J.; Bruhn, P.R.; Gouzoules, F.; Lee, S-H. *J. Chem. Soc., Chem. Comm.* **1989**, 659.

55. Pearson, A.J.; Bruhn, P.R. *J. Org. Chem.* **1991**, *56*, 7092.

Index

cycloisomerization, using palladium(II)
    catalyst, 77
    intramolecular, cyclization, 117
epi-ß-santalene, synthesis, via η³-allylnickel
halides, 299
epoxides:
    reaction with Fischer carbene complexes,
        154- 155
    ring opening, with lithium organocuprates,
        67
esperamycin, antitumor agent, 101
esters:
    conjugated,
        1,4-addition with dilithium dicyano
            dialkylcuprate, 70
        alkyl substituted, diorganocuprate 1,4-
            addition, 68
        diorganocuprate 1,4-addition, 68
        from chromium carbene complex photolysis,
            178
    hydrogenation, homogeneous catalyst, 48-49
    reactivity with lithium diorganocuprates, 67
    reduction, using iron hydride, 61
    synthesis, insertion, η³-allyl palladium,
        287-288
    ß,G-unsaturated ester, from carbonylation
        reaction, 135-136
    Wacker process, for ß-keto compounds,
        204-205
ethers, aryl, from chlorobenzeneiron and phe-
noxides, 328-329
ethyl, migratory aptitude, 29
ethyl benzenes, chromium complexed, α–alkylation, for isopropylbenzene, 320
ethylene:
    palladium(II)-catalyzed "oxidation", to
        acetaldehyde, 203
    Wilkinson's complex poison, 51
exocyclic alkenes, preparation from functionalized vinyl cuprates, 75

F

Fischer carbene complexes:
    coordinatively unsaturated metal(0) d^6, 153
    from metal hexacarbonyl and organolithium,
        151-152
    iron carbene complex analogs, preparation,
        172
    nonoxygen heteroatoms, preparation, 154
    reactions,
        for alkoxy-carbene complexes, 162
        aminocarbenes, 154-155
        bases, 155
        cyclopropanation of electrophilic olefins,
            158-160
        Diels-Alder, 157, 167

dissociative ligand exchange, 153-154
Dötz reaction, 166-171
homologated carbene complexes, 154-155
intramolecular cyclopropanation, 162-164
metathesis, 160-161
Michael addition, 157
oxidation to carboxylic acids, 158
Pauson-Khand process, 157-158
phosphine ligands, 154-155
photochemical activation, 153-154
synthesis of bicyclic compounds, 165
synthesis of quinone derivatives, 167-168
thermal activation, 153-154
to "enolate" anions, 154
fluorobenzenes, ortho-lithiation, preparation of
heterocyclics, 318-319
fluorosulfonates:
    carbonylation, 103
    coupling, by palladium(0) transmetallation,
        98
formylation, of (η⁴-cyclobutadiene)iron tricarbonyl, 225
d^8-four-coordinate unsaturated, square planar
geometry, 13-14
frenolicin:
    palladium-catalyzed alkoxycarbonylation,
        318
    synthesis, 318-319
Friedel-Crafts acylation, of (η⁴-
cyclobutadiene)iron tricarbonyl, 225

G

geraniol, selective reduction, 55-56
glycine, preparation via carbene complexes, 179
Grignard reaction, nickel- or palladium-
catalyzed, 87-88
Grignard reagent, homologated, preparation,
120- 121
bis Grignard reagent, from α,Ω-dienes, with
dicyclopentadienylzirconium(IV) dichloride
catalyst, 116, 120

H

hapticity, 21
Heck reaction, 22, 103-113
    acrylates, 107
    alkenes, 107
    ß-hydrogen elimination, 107, 112
    chiral ligands, 107, 110
    conjugated enones, 107
    electron-rich alkenes, regioselectivity, 107
    enamides, 107
    enol ethers, 107
    examples, 107-112
    intramolecular, 107-108
    mechanism, 106

nitriles, 107
related to "alkylative" heterocyclization of alkynes, 238-239
three carbon-carbon bonds formed, 110-111
heteroatom group, migratory aptitude, 30
heterocyclic:
coupling, Grignard reaction, 89
from zirconacyclopentadiene, 116-117
heterogeneous catalyst, reduction of carbon-carbon double bond, 48
1,3,5-hexatriene, preparation, 79
3-hexenyloxycarbene, to bicyclic cyclopropanes, 164
homoallylic alcohols:
in carbene complexes, cyclopropanation, 162
from η^3-allylnickel halides and carbonyls, 302
selective reduction, high enantioselectivity, 55-56
homogeneous catalyst:
advantages, 48
hydrogenation,
bond specificity, 57
dihydrides, 49-51
chlorotris(triphenylphosphine) rhodium, 49-51
monohydrides, 48-49
hydridocarbonyltris(triphenylphos-phine)-rhodium, 48-49
homogeraniol, selective reduction, 55-56
homopropargyl alcohols/amines, cyclization, 238
Hückel calculation, for η^3-allylpalladium, nucle-ophilic attack, 36
humulene, cyclization, 132
hydrazones, chiral, chromium complexed, for planar chiral benzaldehydes, 325-326
hydrides:
ligand,
in migratory insertion, 29
oxidation state, 3
migratory aptitude, 30
oxidative addition, 23
hydrido, ligand order, 18
hydroboration:
alkynes and alkenes, 92-93
as transmetallation reaction, 82
hydrocarbon activation, concerted oxidative addition, 24
hydrocarboxylation, 135
hydroformylation, 135
using σ–carbon-metals, 76
hydrogen, molecular, in asymmetric reduction, 61
hydrogenation:
asymmetric,

bidentate coordination of olefin, 52
homogeneous, 51
catalysts, 51
ruthenium BINAP catalyst, 54
spectroscopic characterization, 52
see also reduction, asymmetric
catalytic,
concerted oxidative addition, 24
using σ–carbon-metals, 76
homogeneous,
dihydride catalysis, 49-51
monohydride catalysis, 48-49
prochiral olefins, reduction, 51
hydroquinones, from Dötz reaction, 166-167
hydroxyls:
ligand order, 18
metal carbonyl reaction, 32
hydroxymethyls, migratory aptitude, 29
hydrozirconation, olefin reactivity order, 80-81

I
imides, from alkyne-isocyanate cyclocoupling, 252-253
imines:
bicyclic, preparation via cyclodimerization, 122-123
ortho addition, to arenechromium tricarbonyl, 311
photolytic reaction with carbene complexes to give ß-lactams, 244
immobilization, in Pauson-Khand reaction, 244
indane, from Dötz reaction of aminoquinones, 171
2-indanol, racemic, chromium complexed, stere-ospecificity, and products, 323-324
indanone, optically active, from 2-indanol chromium tricarbonyl, 323
indene, "slippage" exchange, 21-22
indium, indium(III) cocatalyst, with conjugated enones, regioselectivity, 293
indoles:
complexed, nucleophilic alkylation, 312
functionalized, synthesis via σ–alkylpalladium(II) complexes, 209
lithiation, via arenechromium tricarbonyl, 4-position product, 318
via zirconabenzyne intermediate, 256-257
intramolecular process, Heck reaction, regiose-lectivity, 107-109
iodides, ligand order, 18
iridium:
cyclooctadienyl iridium(I) complexes, as catalyst, 58
iridium catalyst, "super unsaturated," 58
iridium(I) complexes,
d^6 18-electron saturated complexes, 23

conjugate alkylation, using mixed
cuprates, 70
Wacker process, for ß-keto compounds,
204- 205

L

lactam, and aminocarbenes, for cyclic amino
acids, 178
ß-lactam:
from η^3-allyliron lactam, 137
from intramolecular carbonylation, 104-105
optically active, preparation, from chromium
carbene complexes, 175-176
lactone:
from intramolecular carbonylation, 104
macrocyclic,
coupling, with tin compounds, 96-97
cyclization, 132
via carbonylative coupling, 99-100
reduction, using iron hydride, 61
Γ-lactones, from cyclocarbonylation, 135-136
ß-lactones, from cyclocarbonylation, 135-136
large rings, cyclization, 132
Lewis acid modified cuprates, 1,4-addition,
71-72
ligand:
π-allyl, 6
anionic, 5-7
mononegative, bonding mode, 7
two-electron donor, 6
cationic, 8
bonding mode, 8
classes, 5-11
exchange, 16-22
bulky ligands, rate, 19
coordinatively unsaturated 16-electron d^8
square-planar complexes, 17-18
18-electron saturated system,
dissociative S_N1-like, 18
rate, 18
promotion in 18-electron systems, 21
16-electron d^8 square-planar complexes,
coordinatively unsaturated, 17-18
formal neutral, to formal mononegative
ligand, 39
incoming, order, 18
innocent, 16
and metal oxidation state, 3-4
monohapto, 6
mononegative anionic, bonding mode, 7
neutral, 6-7
bonding mode, 6-7, 7
nucleophilic attack, 31-40
optically active, with organocopper reagents,
72
spectator, 6, 8, 16

substitution, 16-22
1e, associative, 17
2e,
associative, 17
for platinum(II) complexes, 17-18
radical, 17
redox, 17
S_N1-like, 17
S_N2-like, 17
trans effect, 18
transition metal coordinated, electrophilic
attack, 42-44
trihapto, 6
unsaturated,
formal insertion into *cis* metal-ligand
bond, 29
in migratory insertion, 29
π-acceptor, 12
σ–donor, 12
lithiation, arenechromium tricarbonyl, ortho-
lithiation, 317
lithium naphthalenide, reaction with copper
cyanide *bis*(lithium bromide), 73

M

macrocycles:
via intramolecular allylic amination, 275-276
via intramolecular palladium(0) transmetal
lation, 95-96
maleic anhydride, reductive elimination pro-
moter, 28
maleoyl complexes, intermediate in alkyne-
transition metal-carbon monoxide reaction,
254
manganese:
arene metal complexes, stable, 307
arenemanganese tricarbonyl,
alkoxylation, 330
alkylation, 330
arylmanganese(I) complexes,
transmetallation, to palladium(II)
complexes, for olefination, 85-86
"masochistic stereoinduction," 177
mercaptides, reaction stability, 69
mercury, transmetallation, 84
metal(0) d^{10}, with sp^2 halides, agostic bond, 25
metal acyl enolates, 146-148
metal acyls:
alcohol cleavage, 33
amine cleavage, 33
metal alkene complexes, applications, 199-220
metal alkynes, nucleophilic attack, 237-239
metal carbonyls:
alkoxide reaction site, 32
amine reaction site, 32
anionic, strong nucleophile, 137

cationic,
    nucleophilic attack, 135-136
    oxidation, cerium(VI) complexes, 136
    complexed with arene, nucleophilic attack,
        38
    homoleptic, 129-130
        physical properties, 130
    with hydroxide, to metal carboxylic acid, 32
    neutral,
        and nucleophilic attack, 31
        to acyl "ate," 31
    nucleophiles, 129
    organic substrates, 129
    reaction, 32
        coupling, 131-135
        general, 130
    saturated, ligand exchange, 19
metal carboxylic acid, from metal carbonyl and
    hydroxide, 32
metal dienes:
    chirality, 222
    complexation, ligand stabilization, 226
    nucleophilic attack, 226-229
metal dienyls, cationic, nucleophilic attack,
    229-234
metal-assisted carbonylation, 32
metal-hydrogen bond:
    alkene insertion, 30
    alkyne insertion, 31
metallacycles, reactions, 115
metallacyclobutanes:
    as intermediates, 160, 164
    synthesis, 36
metallation, ligand directed, 114-115
metals:
    electron deficiency, and nucleophilic attack,
        31
    electrophilic attack,
        cleavage with inversion, 42
        site, 32
    hapticity, 21
metathesis, 161
methanol, cleavage, in carbonylation reaction,
    135
methyl, migratory aptitude, 29
methyl acyl, anionic, *see also* acyl "ate"
methyl Meerwein's reagent, in acyl "ate" alkyla-
    tion, 152
α–methyl phenyl glycine, synthesis, 309-310
methyl triflates, in acyl "ate" alkylation, 152
4-methylanisole, and cyclohexadienyl iron, reac-
    tion, 231
α–methylene-Γ-butyrolactone, from η³-(2-car-
    boethoxyallyl)nickel bromide and carbonyls,
    302
methylenecyclopropene:

cycloaddition, palladium(0)-catalyzed, 294
    trimethylene methane reactions, 294
methyls, ligand order, 18
Michael addition, of α,ß-unsaturated carbene,
    157
migratory insertion:
    ß-hydride elimination, 30
    mechanism, 29-31
    migratory aptitude, 30
mixed alkyl heterocuprates, for acid chloride to
    ketone conversion, 69
molybdenum:
    η³-allyl molybdenum complexes, 295-296
    stereochemical control, 295-296
    arene metal complexes, stable, 307
    cationic, nucleophilic attack, 36
    CpMo(CO)$_2$+, reactivity increase, 227
    Fischer carbene complexes, 151
    molybdenum acyl enolates, reactions, 148
    molybdenum(II) dienes, proton abstraction,
        to π-allylmolybdenum, 224
    Mo(RNC)$_4$(CO)$_2$, alkylation, efficiency, 295
    neutral η³-allylmolybdenum, 227-228
        conversion to cationic nitrosyl, 228-229
        for cyclohepta-1,3-diene *cis*-1,4-
            difunctionalization, 228
        for diene *cis*-1,4-difunctionalization, 228
monocyclopropanation, dienes, 158-159
Monsanto acetic acid process, 135
R(-)muscone, synthesis using organocopper
    reagent, 72

N
naphthalenes:
    "slippage" exchange, 21-22
    substituted, *trans* difunctionalization, 316
naphthoquinones, from alkynes via cobalt
    phthaloyl, 254-255
Naproxen, anti-inflammatory, synthesis via
    asymmetric hydrogenation, 54-55
neocarzinostatin, antitumor agent, preparation
    using cobalt complexes, 241-242
Nicholas reaction, cobalt-alkyne complexes, 240-
    242
nickel:
    η³-allylnickel bromide dimer, from coupling,
        131
    η³-allylnickel carbonyl, from allylic halide
        coupling, 131
    η³-allylnickel complexes, 298-302
        generation, 298
    η³-allylnickel halides, 298
        allyl group introduction, 300
        chemistry, 298
        radical-chain process, allyl group transfer,
            298

reaction with organic halides, 298-299
  reactivity series, 299
bis-η³-allyl nickel, cyclooligomerization
  reactions, 266
bis(cyclooctadiene)nickel, 298
1,4-*bis*(dimethylphosphine)-n-butane nickel
  chloride, as Grignard catalyst, 88
1,4-*bis*(diphenylphosphine)-n-butane nickel
  chloride, as Grignard catalyst, 88
η³-(2-carboethoxyallyl)nickel bromide,
  reaction with carbonyls, 302
nickel(0) catalyst, cyclooligomerization of 1,
  3-dienes, 265
nickel(0) complexes,
  cycloaddition catalysis, 294
  cyclocoupling of alkynes with
    isocyanates, 252-253
  metallacycles, reactions, 253
  reductive coupling of alkynes, 253
  to nickel(II) complexes, oxidative
    addition, 23
nickel(0) d^{10}, catalyst, cyclodimerization of
  alkenes and alkynes, 122
nickel(0)tetrakis(phosphine) complexes,
  ligand exchange rate, 19
nickel acyl,
  anionic, 142
    acylation reactions, 141
nickel phosphine, Grignard reaction cross-
  coupling, 87-88
nickel tetracarbonyl,
  coupling, allylic halides, 131
  ligand exchange, 19
  properties, 132
  reaction, coupling, allylic halides, 131-135
  reaction with iodides,
    mechanism, 142
    to esters, 142
nitrato, ligand order, 18
nitriles:
  cotrimerization with alkynes, for pyridines,
    249-250
  Heck reaction, 107
  hydrogenation, homogeneous catalyst, 48-49
  ligand, bonding mode, 7
  reactivity with lithium diorganocuprates, 67
  reduction, using iron hydride, 61
nitrito, ligand order, 18
nitrones:
  by rearrangement of O-allyloximes, 216
  in 1,3-dipolar cycloaddition reactions, 216
nitrosyls:
  ligand,
    bonding mode, 8
    spectator, 8
  linear, π-back bonding, longitudinal
    acceptor, 12

nucleophilic attack:
  on ligands, 31-40
  self exchange reaction, 34
  transition metals, 39-40
nucleosides, via palladium(0) catalyzed allylic
  amination, 274-275

O

octadiene, preparation, by reductive dimerization, 262-263
olefins:
  see also alkenes
  catalytic cyclopropanation, with diazo
    compounds, order of reactivity, 183-184
  cis hydroxylation, with osmium tetroxide,
    iron(III) tricarbonyl, 220-221
  disubstituted, Pauson-Khand reaction,
    intermolecular, regioisomer mixture, 247
  electrophilic, cyclopropanation, 158-159
  electrophilic metal complexes, nucleophilic
    attack, 35
  hydrogenation,
    competetion with isomerization, 49
    homogeneous catalyst, 48-49
  ligand,
    in migratory insertion, 29
    order, 18
    π-acceptor, 6
  metathesis, in carbene complexes, 161
  order of reactivity, 183-184
  oxidative addition/insertion, 105-106
  π-back bonding, electrophilic metals, 12-13
  prochiral, asymmetric hydrogenation, 51-54
  tetrasubstituted,
    noncyclic, and Wilkinson's complexes, 51
    reduction catalysts, 58
  trisubstituted, noncyclic, and Wilkinson's
    complexes, 51
organic halides:
  η³-allylnickel halides, reaction mechanism,
    302
  functionalized,
    reactivity, 73
    synthesis of organocopper reagents, 73-74
  oxidative addition, to low valent transition
    metals, 87-88
organic iodides, functionalized, with diethylz-
  inc, 73
organoboron compound:
  from "ortho-lithiation" and trimethyl borate,
    94
  transmetallation, to palladium, 94
organolithium reagent:
  and neutral metal carbonyl, 31
  to generate organocopper reagent, 72
  transmetallation, 91

UNIVERSITY SCIENCE BOOKS
20 Edgehill Road
Mill Valley, CA 94941
FAX: 415-383-3167

ISBN 0-935702-28-8